THE GOLD PROBLEM:
ECONOMIC PERSPECTIVES

The Gold Problem: Economic Perspectives

Proceedings of the World Conference on Gold
held in Rome, 1982

EDITED BY
ALBERTO QUADRIO-CURZIO

OXFORD UNIVERSITY PRESS for the
BANCA NAZIONALE DEL LAVORO and NOMISMA

Oxford University Press, Walton Street, Oxford OX2 6DP
London Glasgow New York Toronto
Delhi Bombay Calcutta Madras Karachi
Kuala Lumpur Singapore Hong Kong Tokyo
Nairobi Dar es Salaam Cape Town
Melbourne Auckland

and associated companies in
Beirut Berlin Ibadan Mexico City

Published in the United States by
Oxford University Press, New York

British Library Cataloguing in Publication Data
World Conference on Gold
The Gold Problem: Economic Perspectives
1. Gold Congresses
I. Title II. Quadrio-Curzio, Alberto
338.4'766922 HD9536.A2
ISBN 0-19-920130-7

Manufactured in the United States of America

NOTES ON CONTRIBUTORS AND ORGANIZING INSTITUTIONS

ROBERT Z. ALIBER	(Professor, Graduate School of Business, University of Chicago, USA)
LOUISE du BOULAY	(Chief Commodity Analyst, Consolidated Gold Fields, London)
† FREDERIC BOYER de la GIRODAY	(Late Director of Monetary Affairs, Commission of the EEC, Bruxelles)
HERBERT J. COYNE	(President, J. Aron & Co. Inc., New York)
LAMBERTO DINI	(General Director, Banca d'Italia, Roma)
GIANANDREA FALCHI	(Officer, Servizio Studi, Banca d'Italia, Roma)
PETER D. FELLS	(Executive Director, Economic Research Branch, Consolidated Gold Fields, London)
DICKRAN GAZMARARIAN	(Managing Director, Mocatta HK Ltd. Hong Kong)
TIMOTHY S. GREEN	(Author of publications on gold, among which, "The World of Gold", London)
ROBERT C. GUY	(Director, N.M. Rothschild & Sons, Ltd., London)
GUIDO R. HANSELMANN	(Executive Vice President, Union Bank of Switzerland, Zurich)
ROY W. JASTRAM	(Professor, School of Business Administration, University of California, Berkeley, USA)
PAUL JEANTY	(Managing Director, Samuel Montagu & Co., London)
PIERRE LANGUETIN	(Vice Chairman of the Governing Board, Swiss National Bank, Berne)
RENE LARRE	(Managing Director, Bank for International Settlements 1971/80; Advisor, Schneider S.A., Paris)
THOMAS R.N. MAIN	(Assistant Manager; Chairman of the Advisory Committee for Economic Affairs, Chamber of Mines of South Africa, Johannesburg)

STEFANO RAINER MASERA	(Director, Servizio Studi, Banca d'Italia, Roma)
NERIO NESI	(President, Banca Nazionale del Lavoro, Rome)
RAYMOND NESSIM	(Group Vice President, Philipp Brothers, New York)
JURG NIEHANS	(Professor of Economics, Berne University, Switzerland)
PETER OPPENHEIMER	(Lecturer in Economics, Oxford University, England)
FRITZ PLASS	(First Vice President, Deutsche Bank AG, Frankfurt)
ROMANO PRODI	(Professor of Economics and Industrial Policy, University of Bologna, Italy)
ALBERTO QUADRIO-CURZIO	(Professor of Economics, Catholic University of Milan, Italy)
ANNA SCHWARTZ	(Staff Director, Gold Commission, US Congress; National Bureau of Economic Research, New York)
LUIGI STELLA	(President, National Federation of Goldsmiths, Manufacturers of Italy, Milan)
PAOLO SYLOS-LABINI	(Professor of Economics, University of Rome, Italy)
FABIO TORBOLI	(President, Goldsmiths' Club of Italy, Milan)
ROGER C. VAN TASSEL	(Professor of Economics; Chairman, Economics Department, Clark University, Worcester, USA)
HENRY WALLICH	(Member of the Board of Governors, Federal Reserve System, Washington D.C.)
THOMAS WEBB	(Head, Balance of Payments and Exchange Markets Division Commission of the EEC, Bruxelles)
GUNTER WITTICH	(Assistant Treasurer, IMF, Washington D.C.)

The *World Conference on Gold*, held in Rome February 5th, 6th and 7th 1982 in Palazzo Barberini and in the Auditorium della Tecnica, was organized by the Banca Nazionale del Lavoro and Nomisma.

BANCA NAZIONALE DEL LAVORO

The Banca Nazionale del Lavoro, a public bank 90% state owned, founded in 1913, is Italy's largest bank. The bank emerged at a national level thanks not only to the absorption of local and inter-provincial credit institutions, but also to the development of its own activity in the sector of specific credits, especially real property, industry, cooperatives and public works. More recently, the Bank has acquired control and shares of a certain consistency in firms engaged in para-banking activities, and has increasingly extended its operations abroad.

NOMISMA

Nomisma (deriving from the Greek word "nomisma" meaning "the real value of things" and later "monetary value") is an Economic Research Institute which was founded in Bologna, Italy, in March 1981. It undertakes research on the domestic and international economic developments of particular relevance for the structure, conduct and performance of the financial and industrial sectors, and carries out research in such areas both independently and as a consultant to Italian and foreign governments, firms and institutions. The activities of Nomisma, directed by a Scientific Committee, are at present organized in the following five divisions: Primary Commodities, Energy, Industrial Trends, Industrial Policies, Sectorial Studies.

CONTENTS

Notes on contributors and organizing institutions v

Forewords
(Nerio Nesi and Romano Prodi) xix

Introduction. Gold problems and a conference on gold

(Alberto Quadrio-Curzio) 1

1. Foreword 1
2. Economic Theory: Views of Some Protagonists 2
3. Economic History: Some Contemporary Evaluations 5
4. Economic Facts since 1968-71 7
   4.1. *Gold as Produced Raw Material* 8
   4.2. *Gold and Inflation* 9
   4.3. *Gold and Oil* 11
   4.4. *Gold as a Reserve Asset with an Unlimited Sovereignty* 12
   4.5. *Gold and the Modified Gresham Law* 14
   4.6. *Gold and the Principal Official Holders* 15
5. Conclusions 17

PART I: MARKETS AND INTERMEDIARIES

Chapter 1. The international gold markets: the view of a Swiss bank

(Guido R. Hanselmann) 21

1. The Current Situation 21
   1.1. *The Situation on the Spot Market* 22
   1.2. *The Situation on the Futures Markets* 23
   1.3. *The Correlation between Spot and Futures Markets* 25
2. The Outlook 27
   2.1. *The Prospects of the Market for Physical Gold* 27
   2.2. *Trends on the Futures Markets* 31

3. Switzerland as Centre for the International Physical Gold Trade 32
3.1. *The Position of Switzerland* 32
3.2. *The Comeback of the Zurich Gold Market* 34
4. The Outlook for the Extensive Remonetization of Gold 35

Chapter 2. The international gold markets: the view of a London banker

(Paul Jeanty) 37

1. Some Introductory Points Regarding Quantities, Worldwide Distribution and Motivations for Gold Purchases 37
2. The "Seven-Year Cycles" in the Gold Markets 38
3. Futures and Physical Markets 40
4. Five Conclusions on the Postwar Period 41

Chapter 3. Physical and futures markets with reference to the US

(Raymond Nessim) 43

1. Fundamental Functions of Physical Gold 43
1.1. *Gold as a Commodity* 43
1.2. *Gold as an International Reserve Asset* 44
1.3. *Gold as an Asset for Private Investment* 45
2. Fundamental Functions of Futures Markets in the US 47
3. Hedging on Futures Markets 48
4. Speculation on Futures Markets 48
5. Other Functions of Futures Markets 49

Chapter 4. The evolution of the German banks' role in the field of gold

(Fritz Plass) 51

1. Relevance of an Analysis on German Banks 51
2. Motives and Prerequisites to Engage in the Gold Market 51
3. Domestic Evolution of the Gold Business 52
4. Relationships with the Biggest Producer: South Africa 54
5. Moving into the International Market and Becoming an Integral Part of It 55

Chapter 5. The Middle East and South American markets

(Timothy Green) 59

1. Middle East Markets 59
   1.1. *Changes in the Markets* 59
   1.2. *Physical Trading* 60
   1.3. *The Investment-Speculation Activity* 61
   1.4. *Central Banks and Government Demand* 61
2. Latin American Markets 62

Chapter 6. Markets and intermediaries in the Far East

(Dickran Gazmararian) 63

1. Growth of Hong Kong as a Major International Market 63
2. The Chinese Gold and Silver Exchange Society 64
3. The Development of the Loco London Market 65
4. The Development of Secondary and Third Markets 66
5. Hong Kong's General Financial Structure: a Fundamental Framework 67
6. Other Far East Centres: Japan, Singapore, Australia 68

Chapter 7. The opportunities and restrictions facing new intermediaries in the international gold market

(Robert Guy) 69

1. The Problem 69
2. Lessons from the Last Decade 69
3. The Strengths of Existing Participants 70
4. Self-Regulation and Official Guidance 72
5. Central Bank Intervention 73
6. New Intermediaries 74

Chapter 8. Markets and intermediaries: conclusions and evaluations

(Lamberto Dini) 75

1. Gold Markets as an Expanding Reality 75

2. The Structure and Behaviour of the Market 76
3. Futures Markets: Stabilization or Destabilization? 76
4. On the Supervisory Role of National Monetary Authorities 77
5. On the Issue of a Return to the Gold Standard 77

(Guido R. Hanselmann) 78

6. On the Supply Side 78
7. On the Demand Side 79
8. Markets Growth in the Future 80

PART II: MEDIUM AND LONG TERM STRUCTURAL ASPECTS

Chapter 9. The medium and long term structural aspects for gold

(Louise Du Boulay and Peter D. Fells) 85

1. Production and Supply of the Market Economies 85
2. Production and Supply of the Communist Bloc 88
3. The Official Sector 93
4. Consumption Demand 94
5. Investment Demand 98
6. Conclusions 99

Chapter 10. Factors affecting production and costs in the medium to long term in the gold mining industry

(Thomas R.N. Main) 101

1. Price Fluctuations 101
2. Price of Gold, Uncertainty and the Mining Industry 103
3. Reasons for Long Term Optimism on Gold 104
4. Planning of Gold Production and Cost Behaviour 105
5. Observations on Countries Other than South Africa 108
6. Conclusions 109

Chapter 11. World monetary demand for gold: key to the gold price outlook

(Herbert J. Coyne) 111

1. Introduction 111
2. Key Factors Influencing Investor Demand 112
3. Central Bank Purchases and Sales 115
4. The Growth of the Private Investor Market 117
5. Successive Emergence of New Markets 117
6. The Development of the Market Infrastructure 119
7. The Increasing Diversity of Gold Media 121

Chapter 12. Gold: Fundamental influences upon the investment climate

(Roger C. Van Tassel) 125

1. Forecast: Sizeable Uncertainty 125
2. The Investment Record of Gold 126
3. Gold's Commodity Role and Long Term Price Trend 128
4. Price of Gold, Inflation and Interest Rates 130
5. Other Influences on the Role of Gold 133

Chapter 13. Aspects of the industrial demand as to jewellery: case study of a typical country

(Luigi Stella and Fabio V. Torboli) 137

1. The Relevance of the Topic 137
2. A Survey of the Italian Gold Jewellery Industry: the Biggest in the World 137
3. Factors for the Italian Leadership 139
4. Prospects for the Gold Jewellery Industry 141

Chapter 14. The golden constant, or the gold standard and the behaviour of commodity price levels

(Roy W. Jastram) 143

1. The Relevance of Statistical Analysis 143

2. Two Hundred Years of English Experience 144
3. Eighty Years of US Experience 146
4. The Stabilizing Evidence of the Classical Gold Standard 147
5. Lessons from the Past and the Present Situation 149

Chapter 15. Inflationary expectations and the price of gold
(Robert Z. Aliber) 151

1. The Problem of the Appropriate Monetary Price for Gold 151
2. The Long Run Equilibrium Price of Gold 152
3. The Market Price of Gold in the Short Run 158
4. Reconciling the Long Run and Equilibrium Prices of Gold 160

Chapter 16. On the instability of raw materials prices and the problem of gold
(Paolo Sylos-Labini) 163

1. Economic Theory and Price Determination 163
2. Price Determination: Raw Materials and Manufactured Products 164
3. Fluctuations of Raw Materials Prices and the Crisis of the International Monetary System 166
4. The Role of Gold 167
5. The World Inflation and the Slowing Down in the Process of Growth 168

Chapter 17. Medium and long term structural aspects: conclusions and evaluations
(Pierre Languetin) 171

1. Gold as a Commodity and as Money 171
2. Supply and Demand 171
3. Before and After 1970 172
4. Two Final Observations 173

(Louise Du Boulay) 173

5. The Supply Aspects of the Market 173
6. The Demand Aspects of the Market 176
7. Final Comments: Objective Analyses and Professional Interests 180

(Robert Z. Aliber) 181

8. Three Questions on Gold 181
9. The Central Issue: Gold and the Monetary Environment 182
10. No Consensus between Economists and Gold Companies 183

PART III: NATIONAL AND SUPRANATIONAL MONETARY AUTHORITIES' POSITION

Chapter 18. Possible and desirable roles for gold in the international monetary system: reconsideration and proposals

(Peter M. Oppenheimer) 187

1. The Historical Background 187
2. Gold, Reserve Currencies and International Payments Equilibrium 189
3. The Position of Gold under Floating Exchange Rates 193
4. A Return to Pegged Rates? 195
5. Appendix A: Milton Gilbert on the Monetary Role of Gold since 1945, by Peter Oppenheimer and Michael Dealtry 197
6. Appendix B: Reserves, Sources and Uses of Gold 202

Chapter 19. The European monetary system and gold

(Frédéric Boyer De La Giroday) 205

1. Introduction 205
2. Gold and the EMS in Its Present Stage 207
   2.1. *The Need for Stable but Adjustable Rates in the EEC* 207
   2.2. *Supporting Mechanism and the Role of Gold in Earlier Monetary Systems* 208

2.3. *The EMS* 210
2.4. *Concluding Remarks on the EMS in Its Present Stage* 216

3. Gold and the Future of the EMS 218
3.1. *Prospects for Changes in Present EMS Mechanics* 218
3.2. *Theoretical Possibilities of Treating Gold in a Future EMS* 218
3.3. *Including Gold in the ECU Basket* 221

4. Conclusions 222
4.1. *On the EMS and Gold* 222
4.2. *On the Wider Framework* 223

5. Appendix: the Inclusion of Gold in an ECU Basket 224
5.1. *Construction of a Basket Containing Gold as well as National Currencies* 224
5.2. *Subsequent Values* 224
5.3. *Composition of the Baskets* 224

Chapter 20. The role of gold in the International Monetary Fund today

(Gunter Wittich) 227

1. Introduction 227
2. Changes in the Role of Gold in the International Monetary System 227
3. The 1975 Consensus on the Future Role of Gold 228
4. The Second Amendment of the Fund's Articles 229
5. Gold in the Fund's Financial Structure and the Gold Sales Program 231
6. Possible Further Use of the Fund's Gold 233
7. The Valuation of Gold and Its Role in World Liquidity 234
8. Conclusions 235

Chapter 21. The past, current and prospective role of gold in the US monetary system

(Anna Schwartz) 237

1. Chronological Changes in US Experience with Gold since 1834 237
2. A Summary of Evidence on the Stability of the Economy when the United States Adhered or Did Not Adhere to a Gold Standard 238

3. The Current Role of Gold in the United States 241
4. Proposals to Increase the Role of Gold in the United States 242

Chapter 22. The role of gold in central banks' reserve composition

(Rainer Stefano Masera and Gianandrea Falchi) 245

1. Introduction and Summary 245
2. Gold in the International Monetary System: 1900-1981 246
3. Gold and Paper Assets: a Risk Return Assessment in the Postwar Perspective 253
4. Conclusions: a Multipolar International Monetary System and the Possible Role of Gold 261

Chapter 23. Obstacles to a return to the gold standard

(Henry C. Wallich) 263

1. A Contemporary View of the Gold Standard 264
2. Three Reasons for Opposing the Gold Standard 264
   2.1. *The Reentry Problem* 265
   2.2. *Post-Reentry Difficulties* 266
   2.3. *Long Run* 267

Chapter 24. Gold operations as an instrument of monetary policy

(Jürg Niehans) 271

1. On the End of Bretton Woods 271
2. Gold from an Investment Point of View 271
3. Gold Operations of Central Banks for Short Term Monetary Policy 272
4. The Present Situation Calls for Some Innovation 274
5. Appendix: the Analytical Basis of the Proposal. A Four Asset Model 274

Chapter 25. National and supranational monetary authorities' position: conclusions and evaluations

(René Larre) 281

1. Peace between Supporters and Opponents of a Monetary Role of Gold 281
2. The Present Situation: High Freedom of Monetary Authorities 282
3. The European Monetary System 282
4. A Glance to the Future: No Remonetization of Gold and Central Banks' Intervention 282

(Peter M. Oppenheimer) 283

5. "It's an Ill Wind that Blows Nobody Any Good" 283
6. Inconclusive Attitudes of Official Bodies on the Use of Gold 285
7. Two Key Factors: Politics and Inflation 285

Index of names 289

FOREWORDS

NERIO NESI
President of the Banca Nazionale del Lavoro

As President of the Banca Nazionale del Lavoro, and on behalf of its Board and its Managing Director, Professor Francesco Bignardi, I am pleased to present this volume.

When Nomisma proposed to the Banca Nazionale del Lavoro the organization of a World Conference on Gold, the subject was not yet deemed to be of current interest; indeed, for many it seemed quite peculiar. But we were struck by two things in the presentation of the proposal. First of all, by the great vitality of gold since 1971 (i.e., from the moment when gold, as a result of the policy of "demonetization", should have begun its decline). Secondly, by the role that a large credit institution could play in order to bring light to bear on a theme and a field of action that economic events had brought brusquely to the fore. Especially in this latter regard, the opportunity arose to offer expert information and material for reflection to State Authorities and Central Banks which, because of their institutional roles, are undoubtedly not in a position to undertake the organization of a world conference on such a delicate and controversial theme.

The results of the Conference certainly provide a significant and long term contribution at a worldwide level and, we hope, represent a reference point for a new policy regarding gold. To this end, the Banca Nazionale del Lavoro intends to promote new initiatives in order that the rich patrimony of the Conference not be lost.

ROMANO PRODI
Chairman of the Scientific Committee of Nomisma

Nomisma's international debut with a Conference on Gold is fully in keeping with its objectives of research, of analysis, of proposals on real economics. It must not be forgotten in fact that "real economics" means looking at the functioning of economic systems both in terms of productive factors and in terms of productive sectors, that is, in terms of the means of production, the processes of production, and the products.

Within this framework a World Conference on Gold finds its precise perspective. This is because gold, as "produced reserve value", is a typical link between real phenomena and monetary phenomena. And it is precisely around this theme that Nomisma decided to organize the Conference.

The worldwide response to the initiative and to its organization around the larger themes of markets and intermediaries, medium and long term structural aspects, national and supranational monetary authorities demonstrates the appropriateness and timeliness of the Programme. We hope therefore that this initiative may represent a constructive contribution to the re-examination of a problem which, in the course of history, has been characterized by a mixture of moods and scientific analysis.

It is my privilege to present this volume on behalf of the Board of Directors of Nomisma and especially on behalf of its Chairman, Professor Francesco Masera, and its Scientific Committee. I further wish to express our thanks to Professor Paolo Baffi for his thoughtful advice. And last but not least, our deepest appreciation to Professor Alberto Quadrio-Curzio, Chairman of the Scientific Committee, whose merit it is to have conceived and planned, within the framework of the activities of Nomisma, the World Conference on Gold.

THE GOLD PROBLEM:
ECONOMIC PERSPECTIVES

INTRODUCTION

GOLD PROBLEMS AND A CONFERENCE ON GOLD

ALBERTO QUADRIO-CURZIO
Chairman of the Conference Scientific Committee

1. Foreword

To introduce this volume of proceedings* of the World Conference on Gold is a task which is simple and complex at the same time. The theme of gold is, in fact, one of those rare topics about which everyone knows something, and many seem to have the first-hand experience to express firm and confident opinions and proposals. History and fantasy, scientific analysis and platitudes, present-day dicta and the beliefs of centuries past overlap and mingle in such a way as to render quite complex both an objective analysis and a detached reading of contributions which seek to be objective.

Thus the aim of the Conference is a difficult one, for the contributions which will surely emerge from the papers presented (and which we in no way intend to anticipate here) will have several meanings and will transmit more than one message. We must, therefore, as we examine the issue, seek in every way to avoid that continual interlacing of sentimental truths, based on facts which are uncertain and doubtful but passionately believed, and factual truths, based on scientific analyses; both are legitimate, provided that they are clearly defined and their limits respected.

This is the scope of the present introduction, while it will be the task of the chairmen of the various sessions to evaluate and draw conclusions regarding the three large themes of the Conference: Markets and Intermediaries; Medium and Long Term Structural Aspects; National and Supranational Monetary Authorities.

When Nomisma decided to organize the World Conference on Gold the issue was still very "suspect": the Gold Commission of the United States had not yet been installed and many dicta were still against gold, even though its not inconsiderable commercial properties (linked however to notable

---

* As Editor of this volume I wish to thank Nicoletta Calabi for her collaboration and Louann Haarman for assistance with the English text.

speculative risks) were recognized, as emerged from debates and conferences on the theme, especially among operators in the field. In other words, the "clandestine" period of gold, which had begun in 1968 and eventually culminated in forms of "intellectual condemnation" of the metal as reserve value and means of payment in international economic relations, was not yet over.

Despite the risk of encountering an uncertain reception, the conviction to proceed with the organization of the conference came from three sources: from economic theory, from economic history, from the economic facts since 1968-71.

2. Economic Theory: Views of Some Protagonists

Even though the problem has been debated for centuries, economic theory has never reached a broadly, or, even less, generally accepted negative judgement on the uses of gold in domestic and international economic relations, either as reserve value, or as numéraire or quasi-numéraire of the economic system. Many great economists have analyzed at length the problem of gold in order to express themselves for or against it. So it has been for the pre-classical economists: Mun, Locke, Petty, Cantillon, Hume, Galiani; and so also for the classical economists, chief among whom, in this regard, was David Ricardo.

Ricardo treated the problem of gold in a series of extremely subtle analyses especially as regards the theory of value. Isolating "an invariable measure of value" was one of his most troubling problems, as, in fact, it has been for all of economic analysis up to the present day. For Ricardo no single commodity can be an "invariable standard" but only an approximation of it. That is, "a commodity produced with such proportions of the two kinds of capital (fixed and circulating) as approach nearest to the average quantity employed in the production of most commodities" (see *On the Principles of Political Economy and Taxation,* 1817-1821, Chapter I, Section VI). He concluded that the single commodity which best approximated the prerequisites of an invariable standard was gold.

Also in more operative terms he dealt with the problem of gold before and during the Bullionist controversy, which evolved in part contemporaneously with the various editions (1809-11) of his noted essay *The High Price of Bullion: a Proof of the Depreciation of Bank Notes.* But his interest in the operative problems connected with gold persisted throughout his life and led him to formulate, and continue to refine, the too little known *Ingot Plan*, which contained — more than a century in advance, as has been recognized by other great economists such as Keynes and Hawtrey — many elements of the gold exchange standard.

Before Ricardo, Smith too treated the problem of gold, as did, with and after him, Malthus, John Stuart Mill, Marx and Cairnes, to name only a few, and also the leading figures of the marginalist tradition, such as Jevons,

Walras, Wieser, Pantaleoni, Marshall, Fisher, Pareto, Wicksell, Cassel. The list is indeed quite long, and might be further lengthened.

But let us come now to John Maynard Keynes, to whose power as a theorist all turn when they wish to demonstrate the absurdity of reflecting upon the uses of gold. How many, however, have actually themselves reflected on the proposals of the great economist? Many cite the famous definition of gold as a "barbarous relic", pronounced in 1923, and consider the argument definitively closed. And in doing so they have already committed a notable omission, because the expression appears in quite a different context: "in truth, the gold standard is already a barbarous relic" ("Alternative Aims in Monetary Policy", Chapter IV of *A Tract on Monetary Reform*, 1923).

And this is not all. It is certain that Keynes did not love gold — so much so that his biographer Roy Harrod wrote, "Affection for the gold standard may yet revive. If it does not, the historian will record that Keynes, almost single-handed, killed that most ancient and venerable institution" (Roy F. Harrod, *The Life of John Maynard Keynes*, Chapter IX, section 2). But it is also certain that his intellectual flexibility did not exclude gold from the possible, useful instruments of economic and monetary policy under given circumstances. Thus, considering the experiences of the gold standard during the 19th century, Keynes wrote "The metal *gold* might not possess all the theoretical advantages of an artificially regulated standard, but it could not be tampered with and had proved reliable in practice" ("The Consequences to Society of Changes in the Value of Money", Chapter I of *A Tract on Monetary Reform*, 1923).

And, even more operatively, having elaborated a plan for the intervention of a single central bank on the gold market, and later a plan for the issuing of international bonds linked with gold, Keynes wrote, in conclusion, two sentences which are too often forgotten, but which are indeed the keystone of his thought and should be, even today, illuminating: "The reader will observe that I retain for gold an important role in our system. As an ultimate safeguard and as a reserve for sudden requirements, no superior medium is yet available. But I urge that it is possible to get the benefit of the advantages of gold without irrevocably binding our legal-tender money to follow blindly all the vagaries of gold and future unforeseeable fluctuations in its real purchasing power" ("Positive Suggestions for the Future Regulation of Money", Chapter V of *A Tract on Monetary Reform*, 1923). "It may seem odd that I, who have lately described gold as 'a barbarous relic', should be discovered as an advocate of such a policy [i.e. of international notes linked to gold] at a time when the orthodox authorities of this country are laying down conditions for our return to gold which they must know to be impossible of fulfillment. It may be that, never having loved gold, I am not so subject to disillusion. But, mainly, it is because I believe that gold has received such a gruelling that conditions might now be laid down for its future management, which would not have been acceptable otherwise" (see "The International Note Issue and the Gold Standard" in *The Means to Prosperity*, 1933). The analytical reasoning of Keynes on gold was not therefore linear, and more than once he recognized, in connection with new projects, its possibile uses.

Scientific analysis regarding gold continued, though at a slower pace, until the great debate of the 1960's on what were subsequently to become the two key proposals: an increase in the price of gold and the maintenance of its institutional role in the international monetary system; the elimination of all forms of convertibility and, with the process of expulsion of gold from the international monetary system, the start of "demonetization".

As we know, history has chosen the second alternative with the institution of the two tier gold market in 1968, the cessation of all dollar-gold convertibility in 1971, and the formal adoption of flexible exchange rates in 1973. But we must not forget that by no means were all of economic theory's illustrious representatives in favour of such a solution. How can we forget the contributions, from different quarters, of Triffin and Gilbert, of Rueff and Harrod and also of some of the participants in this Conference? How can we forget Rueff's support of the proposal to increase the price of gold in order to return to the gold standard, and Harrod's support of the same proposal in order to provide adequate gold reserves and thus protect the working of the system of monetary reserves?

It was, then, the facts, and political choices, and not the theory, to exclude the use of gold in the international (and domestic) monetary systems. Emblematic in this respect are the words which Lord Robbins pronounced in June 1967, presiding over an international conference on the theme of "Monetary Reform and the Price of Gold: Alternative Approaches", with the participation of some of the world's greatest experts in the field. Robbins spoke as follows: "Let me say at the outset that I find many of the arguments which have been directed against this proposal [i.e., that of the increase of the official price of gold] not a little unconvincing. Thus I do not think that we rise to the level of argument appropriate to these momentous questions if we deny the importance of gold as such, or suggest that all arguments for a gold standard system rest upon superstition or psychopathological fixations. This particular mode of attack is surely pretty *vieux jeu* nowadays. The theory of the gold standard — Hume's theory of specie-flow adjustment, Ricardo's theory of the distribution of the precious metals — may have been incomplete, but it was not just a meaningless abracadabra: it was a highly sophisticated analysis designed to show how a world, divided politically into independent units, might yet enjoy unity in respect of international economic transactions" (see Randall Hinshaw, ed., *Monetary Reform and the Price of Gold*, 1967, Chapter II).

With all this we do not wish in any way to support a return to some form of a gold exchange standard, but simply to recall that it was not economic theory to close the door on such forms, but rather a general "state of mind" or "mood" deriving from specific choices. It is therefore with great satisfaction for the long term objectivity of economic science that we now see re-emerging theoretical contributions on the argument, of which we have some distinguished examples also in the papers presented here. The hope, not for or against gold, but for renewed reflection on gold, turns therefore now to the theory of political economy and economic policy, counting on its objectivity — to which, with this book, we hope to have contributed.

3. Economic History: Some Contemporary Evaluations

The second factor supporting our decision to proceed with the Conference derives from economic history. It was again Lord Robbins in June 1967, who synthesized the problem remarking that "if we turn from theory to practice, who shall say that on the whole paper systems have proved superior to gold? Clearly, whatever hopes we may have for the future — and later on I shall disclose divergence from the support of gold theory in this respect — if we are honest we must admit that until now — with the exception of two deflation periods where the gold parity had been wrongly fixed — the record of paper systems is vastly inferior, as regards both internal and external stability" (see R. Hinshaw, op. cit.).

Supporting this affirmation are the contributions of two scholars participating in the Conference, Roy Jastram and Anna Schwartz. We refer in particular to theses also expressed prior to the Conference, although we shall draw freely here from the texts of their present contributions.

The research of Roy Jastram has indicated the stabilizing power of the gold standard on the monetary system in the two most prolonged and almost continuous experiences of the mechanism: the English experience, which spanned two centuries from 1717 to 1914, and the American experience, from 1834 to 1914.

Considering the English experience, Jastram finds that commodity prices fluctuate along a level plane centering on the fixed price of gold, and that the fluctuations have been such as to compensate one another so that their average is very close to zero. In turn, the purchasing power of gold was not affected by any secular trend for 200 years: in 1717 an ounce of gold bought more or less what it bought in 1914.

The period considered was certainly not for England a period of tranquillity. The country passed from an agropastural economy to an industrial and financial economy of worldwide dimensions. And yet, during this violent evolution, the gold standard functioned regularly generating, in the long term, compensatory movements in commodity prices. An analogous result was obtained in the United States in the period from 1834 to 1914, excluding the long interruption of the gold standard from 1862 to

1878.

Jastram's conclusion therefore is that, notwithstanding the considerable fluctuations in commodity prices on either side of the level line of the price of gold, these were compensatory both in England and in the United States, giving rise to a 350-year trend of long term stability in commodity prices and the purchasing power of gold. Such a result is implicit in the classical theory of the gold standard. But the empirical evidence which verifies it is of not inconsiderable importance.

Anna Schwartz arrives at a similar conclusion in her study of the American experience between 1834 and 1981. She recalls first of all that, with the exception of brief interruptions ranging from a few weeks to a year, there have been only two prolonged periods when the United States did not adhere to some form of the gold standard: the 17-year period including the Civil War (1862-1878) and the period from 1968 to the present.

The conclusion of her analysis of this period is that the gold standard guaranteed price predictability in the long but not in the short term. Deflationary and inflationary periods under the gold standard prior to World War I were in fact corrected in the long run, and predictably corrected, by variations in the production of gold and by the passage of gold from non-monetary to monetary use, encouraged or discouraged by the increase or decrease in the purchasing power of gold. An increase in the purchasing power of gold in a deflationary period determined an increase in production. Thus the stock of money and the general price level increased, and the purchasing power of gold decreased. A fall in the purchasing power of gold determined the opposite chain of events. In such a way, under the gold standard, the level of prices tended to remain stable in the long term, providing thus predictable fluctuations in the purchasing power of money.

After World War I the discipline of the gold standard came to be considered as an impediment to achieving the objectives of growth and high employment through the management of the economy. The serious depressions of the years between the two wars were considered to be the measure of failure of an economic system under the gold standard. The loosening of the link with gold after World War I and its abandonment 50 years later reduced long term price predictability.

These broad analyses of economic history do not lead the two scholars to propose a return to forms of the gold standard, although neither of the two excludes that such a return might occur and may even prove useful. Jastram remarks: "There is considerable interest at the present time in moving to some form of gold standard to recapture a degree of stability in the price level that has not been experienced for some 50 years. Opponents assert that the economic and social context has changed greatly since the time of the classical gold standard; that whatever the lessons learned then, they are not applicable now. But the economic and social circumstances in England changed hugely from the early 18th century until the end of the 19th and into the 20th. And in the United States the decades leading up to World War I were vastly different from the times when that nation's economy was in early consolidation.

"Surely we should not reject out-of-hand the power of a gold standard to establish once again a long run stability of commodity price levels. Especially when we now have our present advantages in the degree of sophistication in finance and the benefits of high technology in near-instant communication" (see Chapter 14, section 5 of this book).

And Anna Schwartz: "The political objective of returning to the gold standard was achieved in the 19th century case, despite opposition from silver and paper money advocates. Whether that political objective is currently achievable cannot be determined from a retrospective view" (see Chapter 21, section 2 of this book). "My conclusion is that advocacy of a role for gold in the US domestic monetary system will increase if there is not convincing evidence during this year and 1983 that persistent inflation has been curbed. The silver issue dominated the US political scene in the last quarter of the 19th century. The gold issue may come to dominate the US political scene in the remaining decades of this century unless the managers of economic policy demonstrate their will to eliminate inflation" (see Chapter 21, section 4 of this book).

We are aware that others can offer, in opposition to these arguments, negative results of the history of the gold standard, and even the two authors cited have indicated many, in particular in the deep deflationary periods. It is certainly not our intention to deny these, but rather to point out that the question of the costs and benefits of various forms of a gold standard is — as far as economic history is concerned — still open. And if we have emphasized more its advantages, it is in order to balance common opinion, which dwells principally on its disadvantages.

4. Economic Facts since 1968-71

The third factor supporting our decision to promote a World Conference on Gold came from recent facts and events, facts and events which take on even more weight insofar as neither economic theory nor economic history has passed widespread negative judgement on the roles and uses of gold. It is thus the strength of the facts, and of the underlying causes, which in the present world scenario can justify renewed reflection on the principles at the basis of the problem of gold.

Slightly more than 10 years have passed since August 1971, when the era of panic on world monetary and financial markets began and all dollar-gold convertibility (already restricted, since 1968, with the institution of the two tier gold market) ceased completely. The supremacy of rank, honour and jurisdiction passed to oil, and, in a world economy where much broader structural ties were emerging which required greater attention to real factors, it was sought in vain to control its strength mainly through financial engineering, special drawing rights, the substitution account and various other instruments. For the official history of gold a "clandestine" period thus began, as a consequence of the attempt to exclude it from international economic relations. Coherently with this political line, the United States and the International Monetary Fund aimed, through massive sales, among other things, at a "demonetization" of gold. The "factual" history of gold was instead quite the opposite, with the emergence of a solid vitality sustained, notwithstanding occasionally violent price fluctuations, by certain structural tendencies which we shall now consider briefly.

4.1. *Gold as Produced Raw Material*

Gold is a homogeneous, fractionable, liquid and historically accepted reserve value. But it is also a raw material which has limited producibility and which is utilized as a commodity. As such it is the partial result of a productive process and incorporates a certain cost of production. Such properties have induced some economists (see above all Ricardo) to consider gold as a good approximation to the "standard numéraire" and to underline its role as a link between real factors (productive aspects) and monetary factors (liquidity and reserve value aspects). When its function as monetary standard is removed, the limited producibility of gold permits one to foresee a long term increase in its price, even in real terms.

It might be thought that such an affirmation is quite weak since, as far as supply is concerned, the presence of a gold stock of about 80-90,000 tons against an annual flow of 1000 makes the importance of the latter, and the production costs incorporated therein, irrelevant for the fixing of the price, which is instead mainly influenced by movements of the stock. These, in turn, are extremely sensitive to a very complex demand, including industrial, institutional, investment, hoarding, precautional and speculative components with increasing (and at times quite high) degrees of instability.

These considerations are well-founded especially in periods when (as has happened in the recent past) the stock and speculative demand move significantly. But these are surely not normal periods. The long term trend remains therefore sustained by increasing production costs, not only as a result of the general trend of industrial costs, but also because of the exploitation of marginal deposits and the more or less general conviction that production levels should decrease toward the end of the century, even allowing for an increase in prices of gold.

On the contrary, it is clear that prediction regarding the course of demand and variations in its structure proves to be essential in order to evaluate short and medium term price fluctuations, and this, among other things, because changes in the structure of demand may generate supply drawn from accumulated stocks.

The prevailing hypothesis is that industrial demand (especially for jewellery), institutional (central bank) demand, and some of the hoarding-precautional demand on the part of private operators on fractionated gold units (coins, medallions, small bars) should move regularly and have stabilizing effects, increasing in periods when prices fall and decreasing when prices move sharply upwards.

It is much more difficult, indeed impossible, to predict the course of investment demand and speculative demand. These demands played a crucial role in the hysterical price fluctuations in 1979-80-81, and have supported the growth — alongside the traditional physical markets — of futures markets, particularly well developed in the United States, whose strength is now recognized as determinant in the formation of short term prices. Suffice it to say that in 1981 the trading volume was approximately 40,000 tons with market warehouse stocks fluctuating between 70 and 200 tons and with the annual production at about 1000; from 1974 to 1981 the volume was roughly 130-140,000 tons against an estimated total stock extracted during the course of history of 80-90,000 tons.

In any case, however, it is certain that investment, speculative and, to a lesser degree, the other demands have been sustained, in different ways, by another strong structural tendency of the last 10-12 years: inflation. And price fluctuation in the medium and long term, therefore, will depend in no small measure on this.

4.2. *Gold and Inflation*

The relationship between gold and inflation may be considered from two points of view: that of whether investment in gold has compensated inflation, and that of how inflation influences the prices of gold. We are not concerned here with the problem of whether the re-introduction of forms of gold standards can reduce inflation. We have in fact already considered this from the perspective of economic history, but this is not an issue arising from recent facts.

Investment in gold as a powerful hedge against inflation is confirmed by several different analyses. Aliber, for example, has calculated that the annual real rate of return (i.e. the monetary rate of return deflated by the average annual rate of increase of the price level in the country considered) on an investment in gold has been positive, measured both in US dollars and in Swiss francs, in the periods 1965-75, 1965-82, 1975-82. In particular, from 1965 to 1982 the annual real return in dollars was 7.9 percent and in Swiss francs 4.8 percent.

Van Tassel has also shown that in the United States the average monthly return on gold has clearly surpassed both the rate of increase in consumer prices as well as that of various alternative forms of investment in stocks and bonds. And it has done so in periods starting with the beginning of the present era of gold (i.e. between 1968 and 1971), until December 1977 (shortly before the first great leap), until the peak (January 1980), until more recent moments (May 1981).

This does not mean that it has been the best investment in absolute terms. Masera and Falchi show in fact that in the period from 1950 to 1981, comparing investment in SDRs, dollars, yen, German marks, French francs, pound sterling and gold, the last ranked second to the yen in terms of nominal and real returns. In nominal terms, measured in dollars, an initial investment in gold increased more than tenfold, and in real terms 170 percent. But in the period 1971-80 gold holds first position with a nominal increase of 1282 percent and a real increase of 268 percent, as may be calculated from these authors' data.

Nor does it mean that there have not been disastrous losses on the part of those who bought at peak prices. But it does mean that since the period of great disorder in world monetary markets began and the inflationary process exploded, i.e., since 1971, gold has more than compensated inflation. One figure, indeed, would be sufficient in order to reach this conclusion: the annual average compound rate of the increase of the dollar price of gold calculated on the annual average prices was, from 1971 to 1981, 27.37 percent.

And these are facts which, having been intuitively or analytically grasped by large and small holders of monetary capital, provoked the development of a demand for gold as a defence against inflation, either in act or expected. The demand which resulted — showing varying degrees of stability depending upon the different protagonists: investors, speculators, hoarders — generated an increase in the price of gold, and therefore made a notable contribution to the realization of expectations regarding gold as a powerful anti-inflationary device. In other words, inflation served and serves as a primer for demands which proceed then violently and which accelerate the price of gold with respect to inflation itself.

This must be kept in mind in order to explain the absurd increase in the price with respect to the rate of inflation in the second part of 1979 and 1980. Some maintain that the successive sharp drops in the price of gold in 1980-81 determined not only the interruption, but indeed the rupture of any correlation between the price of gold and inflation. We do not believe it to be true.

First of all, one must consider that the upward and downward price fluctuations in the second part of 1979, in 1980 and in 1981 resulted from speculative and politico-military factors which greatly disturbed, but did not cancel out, "normal" rules of growth in which the primer and the inflationary component remain.

In the second place, it must be noted that the scenario in which the redimensioning of the price of gold occurred, after such excesses, is that of a world economy pervaded by something quite new: the widespread conviction, especially among governing bodies and economists, that inflation can be controlled with the policies of monetary restriction presently in force.

This may be the opinion of many people, but we believe that what is guiding operators in this weak period for gold is not so much a long term vision, but rather a vision of short term investment based on the level of real interest rates. The increase in the interest rates on the dollar to nominal levels which exceed 20 percent and to real levels which exceed 7 percent have been historical events and formidable incentives not to buy gold and, to a lesser extent, to "demobilize" it. So the turning point for the price of gold will depend on a substantial downward trend of real interest rates to a level which turns investors' attention from the short to the long term, thus bringing the demand back to gold as a defence against inflation.

It is not easy to say when this critical level will be reached or what it will be. Much depends on the success of the anti-inflationary policies now in force and in particular on that of the US administration. If however inflation is a structural phenomenon of this phase in the development of the world economy and is rectifiable only in the long term — the hypothesis which we continue to prefer — then gold may soon enter a third great cycle of growth which has been foreseen by many, especially in the case of an eventual recovery of oil prices.

4.3. *Gold and Oil*

This is a structural relationship of considerable importance in the sense that the two markets are interrelated even if there is not an ascertained, quantitative link between the two prices, which have had greater similarity in periods when the price of oil increases and none when this price is stationary.

This assymmetry is quite comprehensible because when oil prices increase, not only does there appear a growing demand for gold which is linked to the surplus of the oil-producing countries, but there also arises an external demand linked to inflation in act and expected, generated by the increase in the price of oil. On the other hand, in periods when oil prices are stationary — which correspond to periods of recession in the international economy — the demand for gold falls.

It is clear therefore that if the oil market underwent a structural inversion, the consequences for the price of gold could be notable. But, for the moment, in spite of the weakening of prices in 1981, this does not seem of a lasting nature. The most synthetic conclusion in this regard seems that which appeared in the June 1981 Annual Report of the BIS: "In spite of the emergence of a healthy trend toward greater energetic efficiency and reduced oil consumption... the global oil deficit will remain high".

If this is the case, the effects on the demand for gold will not be lacking for reasons which were explicitly stated in June 1981, by the Minister of an oil-exporting country which has been particularly active in the market of the metal as a reaction to another structural tendency which we shall now consider.

4.4. *Gold as a Reserve Asset with an Unlimited Sovereignty*

Countries which possess gold need not fear decisions of foreign governments which might directly alter the value of gold or condition its use, as may occur with any national or international currency. Of course, such alterations may derive from market forces, but these are more neutral and, in periods of politico-military crisis, which render paper currencies insecure, they push prices upward.

It is not acute periods of politico-military tension which sustain the role of reserve asset with unlimited sovereignty. Such periods, which unleash formidable and hysterical demands, may only cause speculative thrusts upward when the basic economic conditions so permit, as was demonstrated in 1979 and 1980; but they are not capable of inverting downward tendencies of gold determined by economic factors, as was demonstrated in 1981. What economic and politico-military tensions always reinforce is the conviction that the 1980's will be chronically rich with crises generated by the structural ties of the world economy and by the global redistribution of wealth linked to the dislocation of oil. In such conditions it is better not to get rid of a commodity like gold, relying only upon promises for payment, i.e., on paper currencies.

The fact, moreover, that gold is a reserve value which is not based on a debit-credit relationship removes from its holder the risks of insolvency inherent in an international banking system which today appears rather precarious.

It is precisely such properties of gold which can still, in the future, determine purchases on the part of those countries with fewer official reserves — countries which tend to be, in fact, those which for political reasons are most susceptible to the limited sovereignty of paper currencies of international reserves. This is true for the oil exporting countries, which according to the IMF data hold only 4 percent of the official world reserves of gold, for the developing countries, which hold 10 percent (and also, though for somewhat different reasons, for the socialist countries, among which a new protagonist, China, has now appeared).

Such a line of action has even been explicitly stated: for example, it seems that a Minister of an oil-exporting country which was noted for having bought gold in 1980 at an average price of approximately $600 an ounce, apparently answered criticism by pointing out that the purchasing policy of his country was not speculative but protective in that it aimed to increase gold reserves judged at that time to represent too small a proportion of the total reserves.

This type of statement is all the more comprehensible if one considers that the industrialized nations hold 69 percent of the official world gold reserves, plus those of EMS which represent 7 percent of official world reserves and which are "deposited" to the EMCF.

We are certainly not faced, therefore, with a definitive distribution. Given that the total world reserves should be about 80-90,000 tons; that the approximately 36,000 tons of official reserves will certainly not be redistributed; and that the annual production of 1000 tons represents only slightly more than 1 percent of the stock, the action of countries with smaller stocks will center on the 44-54,000 tons of unofficial reserves. Drops in prices can thus be crucial for reallocations.

For some — and at times these include quite authoritative personalities — it is an error to grant gold unlimited sovereignty in as much as its sovereignty is in fact conditioned by the two great producers, South Africa and the Soviet Union. This affirmation is weak because the flow of production in these countries is less than 1 percent of the world stock, and because they are so in need of regular exportation of the metal that they cannot afford to disturb the market. Of course, such countries benefit from the sale of gold. But, as Lord Robbins remarked (see Randall Hinshaw, op. cit.), if an economic policy for gold proved to be advantageous for the entire world economy, it would be absurd not to activate it only because it might benefit two countries which do not enjoy universal popularity.

And these are, paradoxically, the two countries which more than any others are experimenting the unlimited sovereignty of gold with respect to the limited sovereignty of their paper currencies, which have little strength (or are absent, as in the case of the Soviet Union) in world financial markets.

Still, then, today a statement of Ferdinando Galiani paradoxically maintains some validity. Galiani wrote in the treatise *Della Moneta* in 1751:

"Thus it is that a country will never be able to use leather money or *bullettini* [paper money] for a long period. And although paper notes are used in many places for money, still I do not know whether this country which uses paper money, if it became a tributary of a nearby hostile people, I do not know, I say, whether the conquerors would be satisfied to be paid with paper money or whether they would want metal coins. So great is the gap between public faith and common thinking. The latter is as universal as it is unchangeable; the former does not extend beyond those persons and peoples who have reached agreement, and it is liable to be disturbed and often even to be dissolved for the slightest accident: and therefore a people cannot for a long period use only represented money" (see F. Galiani, *Della Moneta*, 1751, Book 1, Chapter 4).

And since gold, even today, belongs to the realities accepted by "common thinking", the question becomes one of establishing whether there can be an economic policy for gold which can be useful to the world economy.

4.5. *Gold and the Modified Gresham Law*

We come to the problem of gold and the modified Gresham law, which sees gold as a privileged official reserve asset and currencies as means of payment. Gresham's law stated that "bad money drives out good". The modified Gresham law states that gold goes into vaults, while paper money circulates.

The history of the last 10 or 12 years is well known: the attempts to exclude gold from the official reserves, the sales of the United States and the IMF, the cancelling of gold from the Statute of the IMF in accordance with a policy against gold. But also well known is how various countries used their gold reserves as collateral in order to obtain international loans, how almost all countries revalued their gold reserves linking them to market prices, how at the base of the ECU there is a gold deposit with a value calculated at market prices, how various central banks of oil producing and developing countries have increased their official and unofficial reserves, how the news ran recently that even central banks of industrialized countries had purchased gold at $400 an ounce.

Beyond these actions there is a crucial fact: the quota of the official reserves of all countries in gold is, at December 1981 prices, 49 percent, and for the industrialized countries it is 59 percent. But these reserves have no operative link whatsoever with domestic and international monetary policy. Thus the problem arises of whether and how such operative rules should be sought in order to tie gold again to monetary policy.

In the papers presented in this volume five lines of thought emerge in this respect, covering the entire spectrum of possibilities.

1) The political and technical undesirability of reactivating gold in domestic and international economic and monetary relations. Gold is accepted as an ultimate reserve, but it is openly stated (Boyer de la Giroday and Wallich) that it should not be touched.

2) The desirability of a return to fixed exchange rates, but the practical impossibility of effecting this until world inflation has been curbed. From this also follows the impossibility of imagining a "return to gold" as a means of regulating the growth of the money supply and reducing inflation, since the only possible causal relationship is the reverse of that proposed: the ending of rapid inflation is a precondition of returning to pegged rates, not a consequence of it (Oppenheimer and, in part, Niehans). Both scholars supporting this line of thought let it be understood that the abandonment between 1968 and 1973 of fixed exchange rates and a link with gold was in all probability an irreversible choice.

3) An interim position of "waiting" regarding the possibility of returning to direct or indirect forms of the gold standard, but the conviction that it has functioned well in the past in historical conditions no less difficult than the present ones (Jastram and Schwartz). The conclusion of the two scholars

supporting this line of thought is that the pressure to return to a link with gold (not a gold standard in the technical sense) in order to regulate the domestic monetary policy of the United States will increase in the future if it does not prove clear to public opinion that inflation may be otherwise controlled.

4) The desirability and feasibility of the purchase and sale of gold by central banks under conditions of flexible exchange rates as an instrument of monetary policy. Such an instrument would mobilize the gold reserves and could be rapidly put into effect for use alongside open market and foreign exchange operations (Niehans).

5) The fifth line of thought, which emerges both from Lamberto Dini and from Masera and Falchi, is that the gold standard is remote and unrepeatable, but the active role of gold in the international financial system (and not its passive role, as someone suggests) has not ended. From a different point of view Sylos-Labini arrives at a stronger conclusion, with a proposal to introduce a revised gold link as an essential instrument in order to reduce speculative movements and inflation, and, obviously, a system of pegged exchange rates with particular rules of convertibility (limited to central banks; in fixed periods; to be activated gradually).

And these latter are types of intervention which we have supported for some time (see for example my recent article, "A Diagram of Gold between Demonetization and Remonetization", *Rivista Internazionale di Scienze Economiche e Commerciali*, Oct.-Nov. 1981).

4.6. *Gold and the Principal Official Holders*

The papers, of course, elaborate all these points in detail. But we are in the realm of facts, and let us stick to them. Two seem the most important: the work of the United States Gold Commission and the possibilities for Europe.

The work of the Gold Commission: one wonders where it will lead*. Certainly to a very high-level scientific and technical analysis of the problem, given the weight and prestige of its members. But we believe that actions undertaken by the government and Congress of the United States will depend much more on the course of the domestic and international economy. There seems in fact to be no doubt that the United States must favour the dollar in international economic relations and support it as much as possible as soon as circumstances so permit. Since the US government must be aware that in order to guarantee gold "coverage" of the world stock of dollars and dollar bonds the price of gold would have to increase enormously (someone says well over any recorded market price), it is clear

---

* When these pages were written the work of the Gold Commission was in progress.

that solutions of this type are unthinkable. In such a case, in fact, the weight of the dollar component of the official reserves of the central banks of the industrialized nations and of Europe in particular would become irrelevant. On the other hand, the fall of gold would heighten the role of the dollar in the official reserves of foreign countries, but at the same time reduce to very little the United States gold reserves, which even that country recognizes to be of the last resort.

The United States is thus faced with a dilemma. On the one hand the recent drops in the price of gold generated by high interest rates, by positive evaluations on the part of the producers (which during the course of 1981 overly concentrated sales), and by European passivity, offer the United States now, paradoxically, the occasion which escaped it from 1971 to 1979 when it launched a policy of demonetization with massive sales "against" gold. It is the chance — and the means are not lacking in such a weak market — to break the price to such a point as to leave a mark for years. On the other hand there is the risk, in following such a course of action, of creating a veritable earthquake on monetary markets and reactions in defence of gold on the part of Europe.

And it is indeed looking at Europe, this other great economic area, that we wish to conclude. With reserves representing 37.26 percent of the world total — 428 million ounces (including those of EMCF and Ecu countries) against the 264 million ounces of the United States — it seems reasonable that Europe should have sought to defend, flexibly, the price of gold in the course of 1980 and 1981. A coordinated plan of purchases — not of large quantities — put into effect a year ago when the price was around $500 an ounce would have probably interrupted the decline in price, not least of all for the "message" thus transmitted.

The effects might have been three: to maintain competitiveness of gold with respect to the dollar and to avoid therefore liquid capital abandoning the former for the latter; to maintain consequently the value in dollars of European gold reserves; to take a step forward towards forms of gold "coverage" of European currencies as a basis for alternatives to a heroic and devastating (for employment and for industry) war of interest rates.

It is not of course certain that such an intervention would have produced the desired results, but it would have indicated a line of policy and perhaps reduced the dimensions of the "losses" that Europe underwent with the fall in the price of gold ($79,442 million less in the dollar value of official reserves from December 1980 to December 1981, with Italy contributing to such losses for a total of $15,455 million).

The conclusion is therefore that if there is an area which at present should concern itself with elaborating a gradual policy for the use of gold in international relations, this is Europe, whose reserves are such as to permit a freedom of movement which is unsurpassed by any other country, including the United States.

The future of gold in the application of the modified Gresham law does not depend therefore so much on South Africa and the Soviet Union, but on Europe and the United States, whose interests may well not even coincide.

5. Conclusions

Without imagining, then, a return to outdated forms of the gold standard, if gold can in part fulfill a stabilizing role, let us not be ashamed to use it, to re-introduce it in the international monetary system from a position of rank — quite close to the position it presently holds — to one of jurisdiction.

The problem is posed therefore by the facts, and it cannot be avoided, even though it is disconcerting to see how little it has been discussed in the last 10 years. These facts have induced us to convene a World Conference on Gold which, through the objectivity of its participants and their roles as principal protagonists (bankers, producers, representatives of central banks and international monetary organizations, scholars, users) might generate lasting reference points. The hope was not, I would like to stress, that of reaching conclusions for or against gold, but rather to call for renewed examination of the issue.

This will not, then, be a conclusion, but a starting point. The course of the future, however, cannot be traced looking only at gold or at the international monetary system, but must consider also — and perhaps above all! — the fact that during the 1970's the structural changes which developed in the world economy of preceding years determined the crisis in the international monetary system. And the backwash of this crisis on the structural changes themselves has produced a tangle of consequences in which it is extremely difficult at this point to separate out causes and effects.

There remain: inflation, a slowing down of growth, and especially uncertainty regarding the instruments of economic policy. This uncertainty finds its roots in the predominant attitude of intervening with "monetary engineering" in a world economic system shaken by structural and real ties.

Among the aims of this Conference, therefore, there has been that of calling attention once again to a reserve value which is a link between "real factors" and "monetary factors".

Part I

MARKETS AND INTERMEDIARIES

Session chaired by Lamberto Dini

CHAPTER 1

THE INTERNATIONAL GOLD MARKETS: THE VIEW OF A SWISS BANK

GUIDO R. HANSELMANN

In my remarks I would like to outline the present situation and the trends developing on the international gold markets. The viewpoint I take will be that of a Swiss big bank which is very active in the international gold dealing business. The division of the gold markets into spot and futures markets, which, as you know, has been in existence for some time, forms a suitable criteria for the vertical structuring and treatment of the subject. Following these more general observations I will then briefly comment on the Zurich Gold Pool.

1. The Current Situation

The gold price is often described as a barometer of the political and economic mood whereby the gold price and the economic and political climate tend to move in opposite directions. The downward trend noted in the gold price since the autumn of 1980 could therefore lead us to conclude that we live in a relatively favourable economic and political environment marked by low inflation rates and the absence of political tensions. Unfortunately, the opposite is all too true. The international political situation is characterized by mounting strain in East-West relations, and the problem of inflation remains largely unsolved despite some impressive initial success. The failure of the gold price to act as a kind of weather vane to indicate which way the economic and political winds are blowing is perhaps best explained by the anomaly of this same economic and political reality. While inflation or at least the inflationary base is relatively high by historic standards, most industrial countries are going through a recessionary phase. The massive rise in unemployment, the dismal economic prospects and the fact that inflation and high budget deficits have narrowed the leeway for economic policy measures designed to stimulate the national economies are all factors creating an atmosphere of fear and uncertainty concerning our economic future. In such a situation one finds that gold loses some of its appeal regardless of the many existing political conflicts throughout the world. The reason for this can be found in the fact that the bullish effect of the political factors on the price of gold is more than offset by the bear effect of the business and social conditions. Hedging against inflation loses in importance because investors prefer high-interest-bearing placements in as liquid a form as possible to a commitment in gold. Although the fear of a recession inhibits investments in tangibles, inflation notwithstanding, this fear has not been strong enough, at least until now, to trigger an actual stampede out of such assets.

Gold's loss of attraction is further reinforced by the atypical prolonged high level of interest rates. The interest rate run-up we have been experiencing for most of the last two years reduces the incentive to invest in gold. This interest effect already had a negative impact on the gold price last year, although the economic outlook was still relatively bright then. Given that the conditions on the spot market differ from those on the futures markets, it seems advisable to analyze these two market segments separately in order to obtain a differentiated and realistic picture of the processes on the international gold markets.

1.1. *The Situation on the Spot Market*

The constellation of demand and supply factors on the spot market contradicts the price performance in some respects. Interestingly enough, the demand for physical gold has not weakened in proportion to the unfavourable economic environment but continues at a relatively high level. The principal cause of this phenomenon is the existence of strong back-log demand by the gold processing industries as an aftermath of the sharp gold price rise in 1980. A brief flashback can illustrate the price fluctuations and their effect on the industrial demand for gold. In 1979, the gold price averaged $ 306 per ounce; in 1980 it doubled to $ 612. In the course of 1981 it fell back again to $ 460 and is presently fluctuating within the price range of $ 360 and $ 390. Industrial demand for gold developed along the exact opposite lines. Back in 1979, industry absorbed close to 32 million ounces, or almost 86 percent of the entire new production of gold, whereas in 1980 industrial purchases were down to 10 million ounces or 32 percent of the new production. Last year, despite the downturn in economic activities, a substantial revival in the demand for physical gold by the processing industries took place with a view to building up inventories. This stock accumulation is continuing, if on a more selective basis and at a slower pace. In addition, investors with long-range strategies and monetary institutions are now making an appearance as buyers of physical gold. This has been coupled for some time with a noticeable increase in demand from Japan. Growing Japanese interest in gold can be attributed, not only to the better performance of that country's economy compared with most other nations but also to the relatively low purchase price for gold expressed in yen. Slowly easing per ounce prices on the one hand and the firm yen exchange rate on the other make the purchase of gold with Japanese currency an attractive proposition. It is estimated that the increase in the Japanese demand for gold, produced by this "price effect", has at times actually more than compensated the decline in demand for interest rate reasons.

This "price effect" also has its reverse side however, that is, the producers of gold have to increase their sales in order to obtain the same total revenues from them. Thanks to the high average gold price of $ 612 that prevailed in 1980, South Africa was able to produce greater revenues with a lower quantity of gold and even add part of its new production to its reserves. In 1981, in contrast, the pronounced fall in the gold price forced South Africa to sell all of its newly mined gold in order to counteract the emerging deficit on current account. Besides, the world's major gold producer had to procure additional foreign exchange holdings by concluding new gold swaps to the tune of some 3 million ounces. By engaging in gold swaps South Africa has gained a certain, if limited, elbow room to escape from the vicious circle of falling prices and rising supplies. The Soviet Union is also confronted with serious balance-of-payments problems. Lower revenues from the sale of oil in response to declining oil prices, heavier needs to import agricultural products due to bad harvests as well as foreign exchange aid to Poland widen Russia's foreign currency gap thereby increasing its need to sell gold. It is safe to assume that the Soviet Union has at least doubled its gold sales in 1981 compared with the preceding year.

Increased supplies of physical gold on the part of the producer countries were nevertheless absorbed by the growing demand of the processing industries, monetary institutions and Japanese buyers. The downtrend in the prices per ounce noted on the markets can therefore not be adequately explained by the developments on the spot market.

1.2. *The Situation on the Futures Markets*

Trading on the futures markets in the United States has been characterized by a sustained bearish trend for some time now. These markets are very strongly influenced by monetary conditions, economic expectations, political factors and similar elements. The monetary climate exerted its most dominating impact on events in the futures markets during the first half of the year. The historically unusually high level of interest rates in the United States generated a lasting bear mood on the futures markets. But quotations also reflected the volatile interest rate fluctuations, in that drops in interest rates produced a price rise in most instances while interest rate jumps led to price drops. Furthermore, the price performance of the futures mirrored not only the interest movements themselves but also expectations and forecasts regarding money supply changes and interest rate developments. These contrary movements — gold prices rise as interest rates decline and vice versa — were especially pronounced in March of last year, when Euro-dollar interest rates for 90-day deposits slumped from more than 17 percent to less than 14 percent whereas the gold price rose from about $470 to more than $550. Conversely, the gold price dropped to below $400 in July when interest rates moved up again to the higher level of between 16.5 percent and 19.5 percent. In the course of the second half of 1981, with its accumulation of negative economic news, and an ever growing number of forecasts painting the picture of a recession on the wall, pessimism about economic developments became the price-determining factor for the futures. Fears of a deepening recession dominated the scene to such an extent that not even the steep interest rate decline from around 19 percent in late August to 11 percent at the end of November managed to produce a noticeable change in the mood on the futures markets.

Futures registered even minor political tremors such as the downing of two Libyan fighter planes by the US Navy. The assassination of Egypt's President Sadat and the fear about a possible invasion of Poland also generated short-lived gold price increases. The absence of dramatic follow-up developments regularly led to profit-taking and the corresponding price drops.

The futures markets differ from the markets for physical gold in several essential respects. The structure alone makes them very different from one another. The spot market tends to exhibit monopolistic features with two suppliers — South Africa and the Soviet Union — dominating the market and only a small number of important buyers present. The futures markets, on the other hand, are distinguished by a high degree of openness and a correspondingly large number of participants or traders. This multitude of market participants is favoured by the modest margin requirements. The majority of the participants are speculators, which reflects the different relationship Americans have to gold. The motives which customarily apply to the acquisition of gold by private parties, such as hoarding and accumulating a nest-egg for emergencies or lean years to come, are largely absent here. On the futures markets gold is traded like any other commodity. Gold futures are also used for interest arbitrage and tax straddles, as well as by the processing industry to hedge their inventories. The gold producers, especially those from South Africa, hardly ever participate in futures trading because they sell the bulk of their gold production directly to the central bank. According to the regulations of the South African Reserve Bank gold producers may sell 10 percent of their production on futures markets, but rarely do so because of liquidity problems which might arise if further margin calls are made.

Other peculiarities of the futures markets are rules and regulations such as the just mentioned margin calls, the related stop-loss orders and the resulting widespread emphasis placed on charts by the market participants. Limitations on price movements through the "limit up" or "limit down" practices should also be mentioned. These market peculiarities can lead to inexplicable autonomous price movements unique by European standards.

1.3. *The Correlation between Spot and Futures Markets*

Ever since the division of the gold market into spot and futures, which unlike 1968 was not introduced by the monetary authorities but resulted from market conditions, the question of the correlation between these two market sectors has assumed central importance. Experience has shown that it is essential even for a banker who is firmly engaged in trading physical gold to keep a close eye on this "paper" gold market. While in the old days one was well advised to stick to the rule that "not all is gold that glitters", one should add today that "gold is not all that glitters", if one wants to avoid losses. This holds true all the more as the volume relationship between the two markets is something like that between a Lilliputian and Gulliver. In 1981 the trading volume on the US futures markets amounted to more than 1.3 billion ounces, which compares with annual new production of roughly 37 million ounces. This means that the annual turnover on the gold futures markets is not less than thirty-five times the yearly new production.

An essential characteristic of the futures markets in the United States consists of the fact that only a fraction of the contracts that are concluded in any way involve physical gold when they fall due. During the last three years only some 1.2 percent of the contracts on the Commodity Exchange in New York called for physical delivery. Most positions are closed by building up a contra position because the speculators, who form the bulk of the participants on the futures markets, are basically only interested in benefiting from price oscillations. The futures markets therefore have only a marginal *direct* influence on the quantity of gold that is bid and asked on the market for physical gold. A certain percentage of the turnover or of the uncovered positions is admittedly secured by actual gold and at the end of December 1981 around 2.37 million ounces of gold were on deposit in warehouses for this purpose. But they made up not more than 6.4 percent of total new production.

However, the futures markets have a major impact on price developments owing to their enormous volume. Let me illustrate this by citing two examples from last year. At the beginning of 1981 the situation on the market for physical gold was characterized by extremely short supplies. Nevertheless the price of gold plummeted from $600 early in January to $460 at the beginning of March. The main factors which were responsible for this price erosion were recessionary expectations, high dollar interest rates and the release of the American hostages in Teheran. Together they exerted a negative influence primarily on futures quotations thereby generating a bear trend which in turn had an adverse effect on the physical market. During the second half of the year, when the producer countries tended to supply the market more generously, the gold price moved up. The uptrend was a consequence mostly of a temporary change in sentiment on the United States futures markets following greater political tension in connection with Poland and the South Africa-Angola conflict, mounting inflation and declining interest levels. The upshot was that the gold price climbed from $390 in the early days of August to $460 in September.

These two examples also show that the futures markets are capable of determining the price trend on the gold market for limited periods and independently of the real market constellation. However, it can be said that by their nature the futures markets only have impact on the short term price trend whereas the physical market determines the long term trend. One can therefore speak of a kind of division of labour. A lasting divergence of the price quotations from the trend on the market for physical gold — which is shaped by long term supply and demand factors — is hardly conceivable in view of the arbitrage possibilities. Should the gold price deviate too much from the long term price trend which is expected to develop, then this differential between the actual and the anticipated price would be exploited by arbitrage which would bring about a correction of the short term price trend.

It is often assumed that the strong influence on the price patterns and the speculative nature of the futures markets have a destabilizing effect on market happenings and tend to produce excessive and erratic price movements. The truth is however, that it is precisely because of the large market volume, their speculative character and their high degree of openness that futures markets are able to smooth the wave of price fluctuations. Speculation has the function of taking over the risk from consumers and producers. This willingness to shoulder the risk at any time is what lends the market its large volume. In a liquid market it is characteristic of speculation to correct excessive price movements which cannot be justified by economic conditions. It has to be borne in mind that as soon as such an inconsistent price picture emerges, a kind of "counter speculation" sets in immediately. This automatic correction of the speculative action therefore has the effect of smoothing price oscillations. Wide price fluctuations, on the other hand, are typical of relatively thin markets with few traders and cause speculators to take to the wings. This correlation can be demonstrated with the help of market happenings in January of 1980. If we take the sum of the "open interest", meaning the uncovered contracts, as the standard measure of speculative fever, it can be noted that this sum developed proportionally opposite to the price trend in January of 1980. In December 1979 the number of open positions rose parallel with the soaring gold price and at the end of the year reached its peak with some 248,000 "open interest" contracts. But the continued rise in the gold price considerably dampened the speculative fever and the number of open contracts dropped in relatively short order to roughly 170,000 in late January 1980. The market volume became thin after the speculators had moved to the sidelines and this thinness encouraged the related price excesses.

It is paradoxical that while speculation is reduced by a kind of "self-elimination" process due to the peculiarities of the futures markets in the United States, price excesses are reinforced by this elimination. This is underlined by the fact that the margin calls led to the liquidation of contracts when prices rose sharply. The increases in margin requirements, a frequent occurrence when markets are volatile, also reduced the number of market participants. The limitations placed on daily price movements by the "limit up" and "limit down" regulations subsequently paralyzed the futures markets for a while. The basic idea of these regulations is to protect the market participants in times of excessive price swings. They were also designed to calm down market events and provide a breathing spell. It seems, however, that the concentration on so much "paper gold" caused real gold and the spot market, which has no such regulations, to be forgotten. Even if the futures markets are paralyzed, trading on the spot markets will continue and the enforced absence of the speculators and the reduced market volume actually encourage price fluctuations.

Nevertheless, all considered the futures markets tend to exert a smoothing influence on price movements and help to even out extremes under normal market conditions. In the final months of 1981 prices could be observed to fluctuate within narrow limits, which can be explained in part by the changed economic environment but to some extent also by the so-called "price-smoothing effect" of the futures markets.

2. The Outlook

After reviewing the current situation on the gold markets the time has come to throw a glance at the both famous and infamous crystal ball in order to find out what trends might develop over the next twelve to eighteen months.

2.1 *The Prospects of the Market for Physical Gold*

Current economic conditions can be expected to restrain activities on the gold market over the near term. Only marginal economic growth in most industrial countries, a further rise in unemployment and the trend towards less inflation are not exactly factors favouring a recovery of the gold price. In such an environment one must, in principle, anticipate that investors' interest in holding gold will decline for a while. However, investors who are pursuing long range investment strategies and monetary authorities should display some demand, especially if interest rates were to drop back significantly and inflationary expectations remain alive. Industrial demand is unlikely to recede despite the negative economic outlook for the near future. There is still a certain back-log demand in industry to replenish the stocks which were heavily depleted in 1980. But this inventory build-up will also be motivated by the expectation that the prices will go up in the long term. It is quite likely, therefore, that fabricating demand in the total demand for physical gold will increase to more than 60 percent in the light of these circumstances. But even this high percentage will still be substantially below the 86 percent acquired by processors in 1979. This structural shift in demand which has taken place since 1979 harbours certain implications for the stability of the gold price. Experience has shown that the higher industry's share in total demand is, the lower price volatility will be, all other conditions being equal.

The market can be seen to be weakened also as a result of the narrowing current account surpluses of the OPEC countries. It should be noted in this context, however, that gold's share in the overall monetary reserves of the OPEC countries is modest by international comparison. While the world average of the gold "quota" was 11 percent at the end of September 1981, it came to less than 2 percent in the case of the OPEC countries. This means that there is a certain back-log demand in this area as well. Some of the OPEC countries could therefore attempt to use a larger portion of the funds available for investment purposes to buy gold; in other words, to restructure their portfolios in favour of gold. This action could prevent declining OPEC surpluses from being reflected in a correspondingly lower volume of gold purchases.

The supply side of the markets for physical gold is likely to be marked by the sustained sales pressure on the part of the producer countries. The Soviet Union may well be forced to step up its gold sales in order to obtain sufficient foreign currency to finance its grain and know-how imports and to support crisis-torn Poland. South Africa will most probably be compelled once more to sell all of its newly mined gold. It is worth noting in this connection that the annual new production volume of gold in South Africa has tended to decline since 1971 and that this has led to less physical gold reaching the market each year. It is estimated that last year new production made up merely 66 percent of the 1970 production volume. In all likelihood, South Africa will acquire the additional foreign currency it needs by means of gold swaps from reserve holdings. But the possibilities to engage in gold swaps are limited for several reasons. For one, the lending bank must dispose of the infrastructural prerequisites because the gold must be delivered to the bank. Not every banking institution meets these requirements. Secondly, the lending bank must have the necessary liquidity at its disposal, something which cannot be taken for granted in times of restrictive monetary policies. Last but not least, above a certain amount the swaps also harbour the danger of causing considerable cluster risks (delcredere and price risks).

Relatively small quantities of physical gold have reached the market until now through forced sales, although sizeable quantities are stored in bullion accounts, some of them bought at very high prices. Gold purchases have been money-losing propositions since December 1979 owing to often massive price slumps, not to mention the loss in interest during the time the gold was held. The exact extent of the loss caused directly by price declines depends on the currency that was employed. For example, the price per ounce of gold eased from $855 on January 21, 1980 to $398.50 on July 8, 1981. During the same time, the kilogram price of gold expressed in Swiss francs dropped from 43,800 Swiss francs to 26,925 francs. A quick calculation shows that the price per ounce in dollars plummeted by 53 percent, whereas the price per kilogram expressed in Swiss francs slipped by only 38.5 percent. The gold price in Swiss francs did not fall back to the level that existed at the end of 1979 until the dollar rate drifted down more recently. If and to what extent these loss positions will be liquidated will depend primarily on the length and severity of the recession. But since we believe that the likelihood of a deepening recession is slim, a real "shakeout" of these positions will hardly materialize. Conversely, this dishoarding potential might be activated during a significant recovery period, thereby restraining the price increases.

Another potential source of supply consists of the gold stocks of those countries which suffer from a sharp disequilibrium between their gold reserve assets and their foreign exchange reserves. These countries may need to sell gold in order to acquire foreign exchange holdings. This is all the more true as we have seen that the possibilities to conduct gold swaps are limited. Portugal could be cited here as an extreme example. Its gold reserves amounted to $7,758 million if we take a gold price of $350 per ounce, while its foreign exchange reserves came to a mere $755 million at the end of September 1981.

All in all, the market for physical gold can be considered to be in a state of extremely fragile equilibrium. Should the general economic environment deteriorate very substantially — a development which we do not consider very likely — then it is possible that the volume of gold offered for sale will upsurge and gather pace like an avalanche whereas demand will fall off and cause prices to tumble. But should the sluggish pace of economic activity persist, without getting dramatically worse, the gold price is likely to move within a relatively narrow range during the current year, more concretely, within a price orbit of $350 and $450. This bearish climate could turn around completely at fairly short notice, however, if a massive rise in unemployment would force economic policy to give way to pressure and to seize the bull by the horns by adopting reflationary measures. But in view of the unwavering stand taken by the central banks to date, such a development is not likely to take place before the end of 1982.

How should we assess the *possibilities of expansion* of the physical gold market? To answer this question we should take a look at gold's traditional functions, namely its use for industrial purposes, for making jewellery, as an investment medium for private individuals and as a reserve asset for national monetary authorities. Although gold's industrial use constitutes the most important function in terms of quantity, it will not be treated here in greater detail because it is safe to assume that it will retain its significance in the future and therefore cannot be expected to furnish any major impulses leading to the expansion of the market.

The picture looks different if we consider the role which gold plays as a medium for private investment. Here physical gold is purchased not only as an investment but also for hedging purposes even though these two motives cannot always be clearly separated. Gold is usually regarded as a kind of financial emergency store or nest-egg for times of political unrest or economic instability, or both. A sort of insurance policy against economic and political disasters. During periods of stability, interest in gold is relatively small. Gold does not earn interest and creates storage problems. Expressed bluntly, gold is a hobby for collectors under such circumstances, but not an investment medium. This situation prevailed during the fifties. Serious interest in gold as an investment instrument emerged only during the sixties in the wake of the currency crises, growing political tensions in many parts of the world and the beginning of the arms race. The demand for gold bullion and gold coins sparked by this interest was reflected by the rise in gold price quotations. In view of the existing economic and political problems worldwide — which we must assume cannot be solved in the short run — the demand for physical gold is likely to increase with the years.

The function of gold as a reserve asset should also gain in importance over the longer term. Regardless of the strenuous efforts of the United States in the seventies to demonetize gold, the yellow metal has successfully defended its position as an official reserve medium. Efforts aimed at the remonetization of gold had their beginnings in the seventies and reached their peak of success in the spring of 1979 when gold was included as a reserve asset in the European Monetary System. In 1980 the official monetary authorities appeared again as net buyers for the first time since 1972. A gold commission appointed by President Reagan in the United States illustrates the change of heart which has taken place in that country. As an official reserve medium gold performs a function similar to that of a private investment instrument. It constitutes the monetary reserve on which to fall back in times of crisis and emergencies. One need only recall the Second World War, when in many cases gold was the only accepted payment medium. But gold is kept as a reserve asset also for reasons of diversification and as a hedge against inflation. Gold is furthermore the only reserve medium which is not based on a debt relationship. As long as the political situation and the economy develop normally, gold too loses its importance as an official reserve asset and becomes the most expensive "luxury" a nation can afford. But the future does not seem to be evolving towards such a well-ordered existence and the need of the national monetary authorities for autonomous official monetary reserves will persist. In line with the expected increase in physical demand for gold, which I have just described, the chances for growth are favourable.

2.2 *Trends on the Futures Markets*

Allow me now to briefly outline the possible development trends on the futures markets. What effects will the general economic environment have on the markets?

Experience has taught us that the futures markets are most heavily influenced by monetary policy and economic conditions, with interest rates and expectations acting as a transmission belt. Applying a restrictive monetary course will produce high interest rate levels. Owing to the direct link between interest costs and the expense of holding gold, gold futures react very sensitively to interest movements. High interest rates therefore generate a bearish trend on the futures markets. Besides, a restrictive monetary policy raises the hope that inflation will recede soon. The presence of high real interest rates then gives rise to fears that the economic downturn will become more severe. These recessionary expectations in turn reinforce the bearish trend. This constellation can be expected to affect the futures for the time being. The other side of the coin is that the sentiment on the futures markets could turn around quite rapidly as soon as any signs of a reflationary monetary policy appear. An excessive expansion of the money supply would both lower interest rates and arouse inflationary expectations, triggering a bullish trend on the futures markets. Since one can assume with some certainty that the FED will continue to keep its monetary reins tight, the bear sentiment on the United States futures markets is likely to continue in evidence this year.

It was mentioned earlier that the futures markets make international gold trading activities highly sensitive to the monetary policy of the United States. Thanks to this transmission mechanism the American monetary authority is unintentionally in a position to contribute to the demonetization of gold. This seems to be more effective than the gold sales by the US Treasury.

Does the future lie in the "futures" or are the "futures" a thing of the past? What are the possibilities of the futures markets to expand? Thanks to their special features and characteristics, the futures still possess considerable growth potential. Futures markets in general and the gold market in particular are an integral part of America's extremely innovative financial system. This system operates like a system of communicating tubes, with capital flowing back and forth among the individual sectors of the economy depending on the economic environment or climate. For example, if interest rates decline, money will flow from the money market to the bond market and into gold futures, or even into both at the same time. The potential of the futures to absorb these flows is unlimited due to the nature of these markets and because no limits have been placed on the creation of paper gold.

The market is moreover distinguished by its high degree of openness, in that trading on the futures markets is promoted by simple rules and regulations. Relatively little capital is required and problems such as transport, storage, insurance and verifying pureness and authenticity are nonexistent. The high degree of liquidity makes it possible to dispose of the liabilities one has assumed at any time. In spite of the tremendous increase in the volume and turnover of American gold futures, they have so far been "discovered" only by a relatively small number of investors, so that a boom might be in the offing if the economic climate is right.

Several reasons seem to indicate, however, that the expansion in gold futures trading will remain restricted largely to the United States. For one, Americans can look back on a long tradition in futures trading, covering not only interest and foreign exchange but also a multitude of commodities ranging from soybeans via pork bellies all the way to precious metals.

Futures trading in gold is only one activity among many. The absence of such futures markets in Europe is not only a consequence of lacking innovativeness but can be explained above all by the difference in mentality between Americans and Europeans. Paying habits are a typical example of this difference. In the United States, cashless payment transfers are the rule, whereas in Switzerland, admittedly an extreme example of European customs, check traffic makes up only an insignificant portion of total payment transactions. Only time will tell to what extent the gold futures market to be opened shortly in London will satisfy a real need. Putting all of these factors into perspective one can conclude in regard to the growth possibilities of both the futures and the spot markets that the quantitative growth potential of the futures markets is by virtue of their nature larger than that of the spot market, but that the former are likely to remain restricted in large measure to the United States for psychological reasons. But if the international political and military tension should increase markedly, the accompanying surge in demand for gold could shift the emphasis in international gold trading from the futures to the spot market.

3. Switzerland as Centre for the International Physical Gold Trade

3.1 *The Position of Switzerland*

After a look at the markets let us now turn our attention to the intermediaries, and especially to Switzerland as a centre for the international trade in physical gold. The significance of Switzerland, and in particular of

Zurich, as an international trading centre in gold has been reduced somewhat as a result of structural changes. In this connection Zurich not only lost some of its importance in the face of the rapidly growing trade in gold futures but also in relation to the other spot markets. During the boom years of the Zurich Gold Pool in the late seventies, roughly 80 percent of the total annual supplies of gold were made available to the international trade via Zurich. Since those days the Zurich share has declined to just under 50 percent.

The reasons for this loss in importance are rooted both in Switzerland and abroad. In recent years, the major producer countries, namely South Africa and the Soviet Union, have demostrated a keener desire to diversify their gold sales in greater measure than was previously the case in order to reduce their dependence on a single buyer. To this must be added the strong increase in the production of Krugerrands, whose share of South Africa's total annual production of gold rose from 0.7 percent in 1970 to roughly 19 percent in 1981.

Only a small portion of the Krugerrand output is sold by the Zurich gold market. Besides, the London fixing's function as quoted price has gained considerably in importance, especially in times of wide price swings, which has helped the London market to secure additional business. The fixing undertaken daily in London enjoys growing popularity especially in the Soviet Union. And finally it should be mentioned that the development of new markets has also helped to weaken the market position of the Zurich Gold Pool. The reintroduction of the turnover tax on physical gold transactions in Switzerland has contributed significantly to this development and the publication — for a time — of the statistics kept by the Swiss customs authorities on the import and export of gold led to a serious loss of confidence in the anonymity of the transactions on the part of both buyers and sellers, who attach great importance to utmost discretion.

But despite this loss in importance Switzerland still occupies the top position in the international gold bullion trade thanks to its major inherent advantages. One of these plus points is the country's long-standing tradition as an international gold centre, which dates back to the thirties. Others are Switzerland's well-known political neutrality, social peace, a stable currency and liberal economic policy, a favourable geographic location and others more. Fundamental importance must also be attached to the financing potential coupled with Switzerland's role as a leading financial centre, its worldwide business links, the know-how of its efficient big banks and the infrastructure they have built up for the refining, storing and transport of gold bullion. These infrastructure facilities are an extremely important advantage which Switzerland has over other leading gold trading centres.

The customers of the Zurich gold trade hail from all parts of the world. An important role is played by the processing industry, especially the jewellery industry. This can be attributed to the fact that Switzerland not only has its own important watch and jewellery industry but also to its proximity to the two largest jewellery manufacturers in Europe, namely Italy and the Federal Republic of Germany. The Swiss banks also occupy a strong position in the Middle and Far East.

The customers who engage in gold trading in Switzerland differ considerably from those on other trading centers, such as London, to name only one example. A Swiss bank has usually entertained business relations with its gold customers over a number of years, and these relations can cover the entire range of the bank's services and facilities, among them portfolio management and money market or foreign currency transactions. This means that diverse and rather close as well as lasting contacts exist between the banks and their clients. In London, on the other hand, gold dealing is conducted through the intermediary of brokers who act as commissioners. Brokers are mainly interested in taking advantage of price differences, whereas in Switzerland the custody and administration of gold holdings also play an important role.

The global networks operated by the Swiss banks allow them to take part in gold trading on a round-the-clock basis. A few of the banks also trade on the futures markets in the United States, but the main purpose of their activities there consists of hedging positions which have been transacted on the spot market on the same day. I might add that there is no real gold futures market in Switzerland. While it is true that forward buying and selling transactions can be concluded in Zurich, the premium for the forward price depends on the corresponding Euromarket rates and not on the demand and supply situation for the related delivery date. For various reasons, some of which have already been mentioned, it appears that the conditions for a futures market à l'Americaine are not given in Switzerland. Aside from the preference of Swiss market participants for physical delivery — a deeply rooted gold mentality — our country also lacks the necessary liquidity because of the virtual nonexistence of a money market in Switzerland.

3.2. *The Comeback of the Zurich Gold Market*

At the beginning of 1982 the ominous turnover tax, which had been imposed on gold purchases two years ago, was eliminated at least as far as gold transactions by central banks are concerned. This measure should significantly strengthen the position of the Zurich Gold Pool in the international physical gold business.

The inclusion of gold among the merchandise subject to the turnover tax led to a noticeable decrease in the trading volume in Switzerland. Although attempts were made to counteract the diversion of business to other markets outside of Switzerland by employing the so-called bullion accounts, it proved impossible to keep central banks from taking their business elsewhere, because some of these institutions are forbidden by their internal regulations to make use of bullion accounts. Their rules require the gold to be stored in the vaults of a major bank, which would have subjected it to the 5.6 percent turnover tax. This "surcharge", as it must be called, prompted numerous central banks and other buyers who did not want collective safeguarding to move to other markets where they could buy gold without having to pay this surcharge. Parallel to this development one could notice a shift in the supply flows to the gold markets which had benefited from this diversion.

Now that the market environment in regard to the fiscal conditions is intact again and as the trading hours of the Zurich Gold Pool are now non-stop from 9 in the morning to 4.30 in the afternoon, the Pool can be expected to recoup part of the market segment it lost to the tax. At a time of growing social, political and military unrest, the traditional advantages and virtues of Switzerland should bear fruit again.

4. The Outlook for the Extensive Remonetization of Gold

In conclusion of my remarks I would like to add a few words on the extensive remonetization of gold — a subject which has cropped up repeatedly in recent weeks and months. The desire to return to a world monetary system based on fixed currency parities coupled with gold convertibility for the dollar or another key currency is, I believe, caused primarily by disappointment and frustration with the insufficient effectiveness of the substitute instruments used since the collapse of the Bretton Woods system. I am referring here to Special Drawing Rights or paper gold, floating, monetarism and the like. Against the background of nervous unrest and turbulence which has engulfed the foreign exchange markets over the last few years, the call for a return to more stability and transparency and the wish to be able to predict exchange rate developments with somewhat greater accuracy are understandable. Experience has taught us, however, that a system based on fixed exchange rates — no matter what its construction — can function only in a politically stable and economically homogeneous environment. But exactly these preconditions are presently lacking in most industrial countries or at least they are not present in sufficient measure. To take the opposite road, namely to reach the goal of sufficient monetary discipline and the harmonization of national economic policy courses by introducing a fixed rate system, has proved to be the wrong track. I think the EMS serves as an example that this would be barking up the wrong tree.

In addition to these fundamental policy obstacles there are some technical and monetary difficulties in the way of introducing a gold exchange standard. The central problem here consists of establishing an official gold price. What price level should be selected, especially as the current prices are still too high due to the built-in inflation bonus, while a lower official price is simply not feasible? Besides, it would be virtually impossible to enforce a fixed official gold price in the event of wide price swings on the free gold market. As these few observations show, the remonetization of gold on a large scale, the return to a gold standard, will remain a dream, at least for the present.

CHAPTER 2

THE INTERNATIONAL GOLD MARKETS: THE VIEW OF A LONDON BANKER

PAUL JEANTY

Any objective analysis of international gold markets and any synthetic point of view on the matter requires some historical perspective. I shall consider here, after a brief introductory note, the temporal cycles of the gold market in the postwar period and the role of the physical and futures markets, keeping in mind that the last 35 years have probably been the most turbulent in gold's history.

1. Some Introductory Points regarding Quantities, Worldwide Distribution and Motivations for Gold Purchases

Of the generally accepted "guestimate" of 100,000 tons of gold stocks in the world today, some 83 percent was produced since 1850, and some 40 percent represents new production since 1946. It is not surprising therefore that gold was considered such a rare metal throughout history and hence its mythical fascination to people at large.

The next point is that of its distribution worldwide.

One area that is well documented is the public sector holdings, even if some central banks or monetary authorities will not or choose not to disclose their holdings to the IMF, and some of them do not belong to it. According therefore to these statistics, reported gold stocks were, at the last count, some 36,000 tons. I would estimate that undisclosed official gold holdings could amount to some 5,000 tons leaving roughly 60,000 tons to the private sector.

Officially disclosed holdings of central banks barely increased in the last 35 years and therefore some 35,000 tons ended up in private hands. This was in marked contrast with the prewar period when, on balance, all of gold production went to swell official holdings. One must also note that in the recession years, but also as a result of the increase of the gold price from $20.67 to $35.00, considerable dishoarding took place first in the East and later on in Europe.

When it comes to the private sector gold, it is very much more difficult to project an image. The French are reported to hold over 5,000 tons in coins or small bars hidden under mattresses. Some estimates double this figure for Indian holdings but this is mainly in the form of ornaments. We do know that over 20,000 tons have been used in jewellery or industry worldwide in the last 20 years, and this would leave some 25,000 tons to cover coins, jewellery and bars held in prior years by private investors and hoarders. This brings us to the problem of demand.

The different types of demand for gold are all important even if their definition is often confusing. One must distinguish, for instance, true industrial demand from jewellery manufacture. The latter must also be separated into two classes — high and low premium jewellery. Last but not least, some distinction must be drawn between hoarding and investment demand. And the various aspects of these types of demand could be even further specified. But this is not the scope of the present paper: other contributions to this Conference will treat the matter in more detail.

The motivations for gold purchases deserve instead some consideration inasmuch as they may be considered to represent the synthetic elements at the base of the market.

I need not dwell on the attractions of gold but must point out that one of its important qualities, its price stability, disappeared in the last ten years. As a result transactions in monetary gold have, to a large extent, been limited to gold swaps, as we shall now see.

Political events and fears of hyper-inflation still remain a primary hoarding motive for individuals. In the last 35 years, however, tax evasion became, and still is, a growing consideration both in Europe and the Far East.

There is a solid base for purchases of high premium jewellery and tradition ensures a good one for low premium jewellery in the East. The latter, however, is very price elastic.

As to investment demand, one can notice a change from a defensive posture for the greater part of the period to a positive approach in the last ten years.

The underlying reason, I believe, in respect of nearly all gold demand is that of inflationary expectations. In the last 5 years short term price fluctuations, particularly reflected by the activity in futures markets, make this very obvious, as we shall see below.

2. The "Seven-Year Cycles" in the Gold Markets

Gold was traded in since time immemorial but until 1939 it was to all intents and purposes but a facet of foreign exchange markets. It is therefore the postwar period that saw major developments in gold markets, their evolution and that of major intermediaries.

As a general rule, major changes were essentially dictated by political decisions, some on an international, some on a domestic basis. Such events appear to follow seven year cycles.

Given this, five periods will be considered:

1) 1947 - 1954: "The First Two Tier Market"

In 1946 the Banco de Mexico appeared as a seller at $40.53. Prices as high as $80,000 ruled in the Bombay market. Prices in Hong Kong were even higher and this was indeed the age of the buccaneers — chartering old bombers to fly gold to the East. The more aggressive bullion dealers and banks, particularly in London and New York, made hay in dealing in "transit" gold.

The desire of the IMF to abolish private gold trading resulted in the creation of several new markets when most traditional intermediaries were prohibited from participating. Some, however, were active indirectly in the growth of "industrial gold" markets and this led to the erosion not only of restrictions but also of the premium over the official price of gold.

When the flow to the Chinese mainland halted Macau, Hong Kong and the Bangkok markets covered the rest of the Far East as well as India. Cairo and Beirut fed Kuwait which, at the time, was the main smuggling market to India. But behind all these relatively new intermediaries, there remained the main players — some London houses, some French and Swiss banks, in fact, the embryo of a future Swiss gold pool.

2) 1954 - 1961: The Re-opening of the London Gold Market

The re-opening of the London gold market provided a very effective medium for both central bank and private transactions.

At the same time the abolition of the turnover taxes in Switzerland allowed the development of a large market for physical gold.

In no time, London regained its prewar position as the recognized centre for international transactions. The main feature was the large transactions handled for account of central banks. Over the ensuing years, such activity represented well over half the turnover.

Growing central bank purchases and investment demand, a large part of which for private American account, created the gold rush of October 1960, which drove prices up to $40.00. Long delayed intervention by the US Treasury finally brought back prices to $35.00 and some disinvestment resulted.

3) 1961 - 1968: The Period of the "Gold Pool"

This was at first successful and the Pool initially increased its holdings. By 1967, after certain defections and currency crises, massive investment and speculative buying by both central banks and individuals resulted in losses of over 2,000 tons to the Pool and led to its demise.

4) 1968 - 1975: "The Second Two Tier Market"

The second two tier market effectively precluded any central bank activity.

The measures taken also provided no floor price for gold producers. As a result the switch of South African sales to the newly established Gold Pool gave Swiss Banks a tremendous advantage over other centres.

Zurich was, for a time, the main world market. However, this position was gradually eroded as London's experience regained its importance and other markets developed worldwide.

5) 1975 - 1982: The "Casino Period"

This is a phase which I would like to describe as the "Casino Period", although this applies to the last three years.

It was decided to cut the last link of gold with currencies. More importantly, however, American gold markets were re-opened and particularly the futures markets were developed.

The re-opening of the American markets proved at first to be a "non-event" and "resulted" in a halving of the price of gold in 1976.

On the other hand, the growth of futures markets, particularly in the United States, was later of paramount importance and culminated in the speculative aberrations of 1979 and 1980. In 1979, encouraged by the low margin requirements and lax prudential rules in the United States, gold speculation became the fashionable game. It snowballed and, aided and abetted by professional and semi-professional manoeuvres, in six months the price shot up from $250.00 to reach $875.00 in January 1980. Twenty four hours later it fell to $585.00.

On reflection, and with the benefit of hindsight, one can only describe such market behaviour as a speculative aberration.

In general the development of futures markets as from 1978, the high leverage and pyramiding of credit such paper markets gave, became of paramount importance in determining the gold price.

3. Futures and Physical Markets

Futures markets are today the major factor in determining prices in the short term and likely to remain so. Their structure and lack of prudential requirements do, however, give rise to some concern.

In the US, it seems strange that the lessons of the very near collapse of major clearing house members — as a result of the Hunt crisis in silver — have not been learnt. If anything, capital at the disposal of the clearing house was considerably reduced by smaller allocations of resources through members' new subsidiaries, many of which are also trading now as principals.

The physical market for gold does, however, demonstrate that it is still the basis for the real demand for gold and will be the major price determinant in the medium term. It is worth considering throughout the period examined not only the elasticity of certain types of demand but also the reactions of such physical markets to speculative excesses.

One should take note of four important waves of investment and speculative buying of gold by private investors in 1960, 1968, 1974 and 1979. They were generally followed by periods of disinvestment and, on the figures available to date, 1981 appears to show private investors disinvestment in excess of 200 tons.

At the same time, the major dishoarding and re-sale of low premium jewellery (in excess of 300 tons in 1980), a reaction to the enormous price increases, only goes to show that the ladies in the East are probably better market judges than professionals and speculators are.

Last but not least, the considerable growth in central bank activity leads one to consider the revival in the importance of monetary gold.

In 1979, the inflationary effects of the enormous increase in the gold price was of considerable concern to a number of central banks. Concerted action to correct speculative excesses or the formation of a new Gold Pool to contain large price fluctuations did not at the time take place but one cannot exclude such a possibility in the future. Price stabilization — not fixing — would be of benefit to both producers, consumers and central banks.

As to the future, it is worth speculating as to the lessons that can be drawn from the periods examined.

4. Five Conclusions on the Postwar Period

In conclusion:

a) The history of the last 35 years goes to show that no international or national legislation can ever stop the insatiable appetite of the world at large for gold. Taxes or restrictions will only divert activity from one market to another and will always be defeated by the ingenuity of major market makers who will temporarily change their base of operation (i.e. Luxembourg is a recent example).

b) International gold markets are today one market operating on a 24 hour basis on a worldwide basis. It can be said that markets in physical gold tend to deal wholesale on a "Loco London" basis but the largest retail market is no doubt in Switzerland.

c) One should also remember that both in spot and futures markets, the trading activities of major Arab money changers and that of Chinese syndicates had in recent years, and still have today, a greater influence than any of the traditional intermediaries.

d) Futures markets will continue to develop and will be of major importance in short term price movements. Another crisis in some commodity markets may be required before both the structure and prudential regulations of American gold futures markets can be corrected by Congressional action. In the meantime, new markets such as the London futures gold market will offer investors greater security.

e) Preliminary figures for the year 1981 show a very strong revival in jewellery demand particularly in the East. This and central bank demand have easily absorbed both large Soviet sales and private disinvestment. One must therefore conclude that despite short term fluctuations, demand for gold will ensure in the medium term a substantially higher price. Inflationary expectations will only accelerate this movement.

CHAPTER 3

PHYSICAL AND FUTURES MARKETS WITH REFERENCE TO THE US

RAYMOND NESSIM

The lure of gold seems to endure. Although the intense interest of speculators has subsided from its peak in early 1980, private demand for gold remains high by historical standards. And today, advisers to the president of the United States are looking to gold as a way to save the dollar from what they see as irresponsible monetary policy.

I would like to speak to you today about the importance of gold markets for the US economy. In particular, I shall address the question of whether speculation in gold is contributing to some of the economic problems of the United States. I would also like to speak to the relationship of the gold futures markets to the trading of gold on the spot markets.

1. Fundamental Functions of Physical Gold

Gold has three functions, which sometimes overlap. It is a commodity that has intrinsic usefulness and appeal and that must be produced. It is an international reserve asset. And it is held privately for speculative investment. To understand the gold markets, we need to consider gold in each of its roles.

1.1 *Gold as a Commodity*

Gold is not important in the United States for its contribution either to the gross national product or to world production. Today, its production makes up less than two hundredths of one percent of the GNP. And although world production is declining, the US contribution in the past decade has been falling even more rapidly, from 8 percent to 3 percent.

But gold as a commodity should not be dismissed. More than 70 percent of the world's private gold purchases in recent years have been for jewellery, electronics, dentistry, and other industrial or decorative purposes. And although gold is but a tiny part of US consumption, it has no close substitutes in a number of its industrial uses. These users need an efficient means of transferring risks that they normally face from unexpected changes in the price. If there is no such mechanism, they will suffer. This is a subject that I will return to later in the talk.

1.2 *Gold as an International Reserve Asset*

I shall next talk about the role of gold as an international reserve asset. I shall not pretend to inform a group of bankers on the changing status of gold in the international monetary system. But let me say a few words, as a participant in the private gold markets, about my understanding of the reserve function of gold.

First of all, the interests of some central banks in holding gold reserves have not died away and appear to be increasing. This is true despite the efforts of the IMF to reduce the role of gold as a reserve asset and despite the strong support of this policy by the United States, at least until recently. I note that in 1980 gold reserves of central banks went up by almost 250 tons, most of which was purchased from new production and the private markets. This was the first time in many years that central banks made large purchases from the private markets. These purchases, combined with a reduction in supply, reduced the amount of gold available for non-monetary use by almost 50 percent. I do not believe that any official measures will succeed in ending or greatly reducing the role of gold as a reserve asset in the near future. Indeed, there is a possibility that the change may be in the opposite direction, if the United States follows the counsel of some of President Reagan's advisers to peg the dollar to gold. You already know the arguments for and against placing a currency on a gold standard. But let me give you a few of my reactions to this proposal in the United States.

First, today's advocates of a gold standard for the dollar often hark back to the gold standard of the late 19th century as evidence of how well it would work today. But the comparison is not well taken. The gold standard of the 19th century was an international system of pegged exchange rates. If the United States were unilaterally to adopt a gold standard, exchange rates between the dollar and other currencies would not be fixed, unless, of course, other countries chose to join the gold standard, either directly, or indirectly by pegging to the dollar.

Apparently, the motivation for pegging the dollar to gold is not a perception that we should balance our external payments with a fixed exchange rate, but rather a desire to have a commodity basis for regulating the US money supply. There is a feeling that central banks have not been responsible in their control of money supplies and that it would be better to link changes and growth in the money supply to changes in the government's holdings of a commodity, such as gold. The hope is that a gold standard will foster slow, steady monetary growth and lower inflation.

The aim is admirable, but a gold standard for the dollar, in my opinion, will not work. It will not work, because the demand and supply for gold are volatile and unpredictable. It is futile to think that a government can set a price that will always reflect what the free market price would be. If the United States were to set the official price of gold too high, compared to what it would be in the free market, gold would be sold in large quantities to the government. As the government paid for these increased gold holdings, the US money supply would increase rapidly, creating the inflationary pressures that the gold standard was supposed to end. On the other hand, if the official price were too low, relative to the free market price, investors would purchase gold in large quantities from the government. This would decrease the money supply and create unwanted deflationary pressures on the economy. Indeed, such purchases could be large enough to deplete US gold reserves, in which case we could no longer maintain a gold standard. If the US stopped selling gold because it ran out of reserves, the price would rise to its market level. The very prospect of such an end to the gold standard creates incentives to buy gold. Since there would be almost no chance of the free market price falling below the official price, buyers would face a one-way gamble. With so little chance of loss, even a remote possibility of the United States losing all its gold reserves could touch off speculative buying that would fulfill the prophecy.

To underline the inability of the governments to set and maintain an official price without intolerable changes in official gold reserves, it bears noting that the United States sold 412 tons of gold in 1979, in an effort to stem the rapid climb in the price of gold. These sales were equal to 40 percent of new production and 5 percent of US gold reserves. But in spite of the size of these US gold sales, speculative demand for gold, and its price, continued to rise. In short, the likelihood of widely fluctuating demands for gold, combined with uncertainties of new supplies from South Africa and the Soviet Union, make it impossible to set a single price that will approximate the free market price for long. In my view, a gold standard for the dollar would most likely lead to less, rather than greater, stability of the US money supply.

Therefore a return to a gold standard without the will of governments to apply financial discipline in their own countries, and without a decision to restrict public expenditures and to limit the growth of money supply will be of no avail.

1.3 *Gold as an Asset for Private Investment*

Let me turn now to gold as an asset for private investment. Only since January 1975, have US citizens legally held gold, except for jewellery or rare coins. There are now a half-dozen or so ways for investing in gold. Investors can purchase bullion bars and coins, open metal accounts with banks or dealers, enter into gold futures contracts, or buy gold options or gold certificates. The great and growing popularity of these speculative investments raises questions in the minds of some whether such investments hurt the US economy. There are concerns that the changeable demands for gold increase currency fluctuations and that investments in gold reduce the level of productive investments.

Like the demand for any asset, demand for gold is related to its current price, the expected return, and the cost of foregoing interest and returns on other assets. The high rates of interest in 1981 have without doubt reduced demand for gold. Gold is affected by changes in interest rates much as any other non-interest-bearing asset. Its special qualities as an investment are found in the expectations about future prices.

When investors are uncertain about dollar assets and about the state of the economy, they often turn to gold. Its price has served as a kind of barometer of economic confidence. Lack of confidence in the dollar relative to other currencies, doubts about the value of currencies in general, and expectations of political instability and accompanying problems for the world economy have all, at times, pushed up the price of gold. Gold is believed to retain its value in the face of failing economies and expansionary monetary policies. The questions are: does such demand for gold *reflect* or *cause* the uncertainties and doubts about other assets? Does the availability of gold as an investment add to the fluctuations in the dollar's exchange rate? Does investment in gold serve as a substitute for, and thus reduce, investment in productive activity? No one really knows the answers to these questions. I do not believe that anyone can demonstrate that investments in gold cause these problems. And there are sensible arguments that they do not, to which I will now turn.

An inverse correlation is often observed between the price of gold and the value of the dollar. But this does not prove that changing demands for gold increase the fluctuations in the exchange rate. When the dollar has fallen in terms of gold and other currencies, the cause has usually been some change in the US economy. At other times, a world political crisis has pushed up the price of gold in terms of most currencies, reflecting doubts about the entire world economy. Rarely is the cause anything specific to the gold market.

The question then is whether investors would refrain from shifting out of dollar assets or out of currencies in general, if investment in gold were not a possibility. I doubt if they would. The removal of the gold market would not eliminate the underlying disturbances. investors would simply choose other commodities, or assets denominated in other currencies, as a substitute for gold.

Next, consider whether investment in gold is a substitute for productive investment in the United States. For individuals, this may well be the case.

But it does not follow that total investment is reduced. That depends upon whether the sellers of gold use their revenue for productive new investments. We can look at the gold market in one sense as a kind of intermediary that provides another asset for investors' portfolios. In this way, gold markets can transfer savings from those who prefer gold to others who now prefer productive investment. Functioning like a financial intermediary, the gold market could actually increase productive investment.

I have talked now about gold in each of its three functions — as a commodity, as an official reserve asset, and as a private investment asset.

2. Fundamental Functions of Futures Markets in the US

I would like to turn now to the role of the futures markets in the United States and how they serve those who buy and sell gold on these markets more efficiently.

A prominent bullion dealer from Frankfurt and a good friend speaking before the Financial Mail's Annual International Conference in 1979 remarked about four reasons why he thought gold could not be traded on the commodity exchanges. One of his four reasons was and I quote:

"I have never yet heard of anybody who wears a couple of Maine Potatoes around his neck and keeps some pork bellies under his pillow, as a reserve of last resort".

I am happy to inform you today that the volume of trading on the futures markets in the US has grown rapidly over the past seven years with close to 13 million contracts traded in 1981, which is equivalent to an approximate volume of 40,000 tons of gold.

Gold futures are traded in the US primarily on the Commodity Exchange in New York and on the International Monetary Market in Chicago.

A well-functioning competitive market requires accurate and immediate price information among distant markets, ease of entry into the markets, sufficient volume to provide depth and liquidity, an absence of credit risk, and opportunities to hedge against unexpected price changes. Futures markets foster and create many of these competitive conditions, not only in the futures markets themselves, but also in the spot markets. Futures markets provide the means for hedging and for price discovery, usually more efficiently than do forward markets.

Futures markets differ from forward markets in several ways. The futures markets provide organized exchanges, with contracts of standard size and with standard delivery dates. Transactions are made anonymously through a central clearing house, which almost eliminates credit risk. Because of these arrangements, the volume of trading for most contracts is large, providing a high degree of liquidity for either long or short positions. Actual physical delivery of commodity, although possible, is unusual in the futures markets.

3. Hedging on Futures Markets

Let me turn first to the possibilities of hedging on futures markets.

Buyers of gold for the manufacture of jewellery, dentistry, or industrial uses face risks of unexpected increases in the price, which may not be matched by increases in the prices of their own finished products. These buyers hedge their risks by purchasing futures contracts in gold, for dates and quantities that approximately match their plans to buy gold. Rather than take delivery on the futures contracts, as they would on a forward contract, they usually sell their futures contracts as the contracts reach maturity, and purchase gold having the exact specifications for their physical requirements in the spot market. If the price of gold has risen more than was expected, their losses in the spot market will be offset by profits on the futures contracts. Producers of gold, on the other hand, face risks of unexpected declines in the price of gold. They could enter the futures market as short hedgers, selling futures contracts at dates that correspond to their anticipated sales of gold on the spot market. If the price of gold does fall more than expected, their losses in the spot market are matched by profits on the futures contracts.

Bullion dealers are also large users of the futures markets, in ways that allow them to serve their customers more efficiently. Whenever the dealer's carrying costs for gold, which include interest, storage, and insurance costs differ from the basis, the dealer can profitably use the futures market. For example, if the basis exceeds the dealer's carrying costs, he buys gold spot and sells a futures contract, effectively borrowing gold and simultaneously lending money to the market. Conversely, if the basis is smaller than his carrying costs, he will sell gold spot and buy a futures contract, effectively lending gold and simultaneously borrowing money from the market. The dealer usually does not hold these contracts until they come due, but cancels his position as soon as the basis comes into line with his carrying costs.

4. Speculation on Futures Markets

Many traders in the gold futures markets never actually hold gold. These are the speculators. Speculators give some people the impression that the futures market is a kind of casino. They are seen as gamblers and are thought to make little contribution to the needs of those who produce and use gold as a commodity. But without the speculators, the futures markets could not function efficiently for the hedgers.

Speculators are usually willing, at a price, to take positions long or short, for any future delivery month, and to assume the risks of price changes.

Because of their willingness to take the opposite positions from hedgers, price fluctuations are usually smoother than they would be in the absence of the futures market.

There is a certain irony about the activities of speculators in the gold markets. On the one hand, speculation in gold is the primary cause for fluctuations in the spot prices. But, on the other hand, the futures markets need the speculators if they are to function effectively and to reduce price fluctuations. Without the speculators the futures market could not provide the liquidity, the price stability, and the ease of taking either long or short positions that make the market efficient for all traders.

5. Other Functions of Futures Markets

In addition to providing opportunities for hedging and helping to smooth price fluctuations, the futures markets are also a source of immediate and accurate price information, which is communicated around the world. This price information greatly benefits the users of the spot markets in gold. As a result, the prices generated by these markets are competitive prices.

Futures markets have proved particularly useful for commodities that are standardized and subject to considerable price volatility. Gold is such a commodity. Gold futures markets play an important role in making the spot market for gold more efficient for all kinds of buyers and sellers.

The influence futures market exerts on the establishment of a world market price for gold cannot be ignored, either by private market operators or by monetary authorities.

It is understandable why futures trading in gold is also being considered in other markets, such as London. In this context, development of the gold options market should also be mentioned.

The Commodity Exchange in New York has applied to the Commodity Futures Exchange Commission for designation as a contract market in options on gold futures. The proposed new contract will be a "Put and Call" option on Comex Gold Futures contract.

Like Gold Futures, gold options serve a need for producers, traders, manufacturers, investors and speculators who have to live and operate in a world of uncertainty, great risk and price volatility.

References

Aliber, Robert Z., *The International Money Game,* Basic Books (3rd Ed.), 1979, pp. 75-79

Bank for International Settlements, *Fifty-First Annual Report*, Basle, 1981, pp. 138- 143.

Bordo, Michael David, "The Classical Gold Standard: Some Lessons for Today", *Federal Reserve of St. Louis Review*, 63, May, 1981, pp. 2-17.

Chicago Mercantile Exchange, "Applications for Contract Market Designation to Trade Futures Contracts in Gold Coins", February 20, 1981 (Table 7, p. 21).

Davies, Jack L., "Gold: A Forward Strategy", *Princeton Essays in International Finance*, No. 75, May 1969, Princeton University: Princeton.

Gray, Roger W., "Price Effects of a Lack of Speculation", *Food Research Institute Studies,* Supplement to Vol. VII, 1967. Reprinted in A.E. Peck, ed., *Selected Writings on Futures Markets,* Vol. II, Board of Trade of the City of Chicago, 1977, pp. 191-207.

International Monetary Fund, *International Financial Survey:* July 28, 1975, p. 214 (US makes gold legal for US citizens). September 15, 1975, pp. 263-265 (IMF reduces role of gold). June 21,1976, pp. 177-182 (IMF reduces role of gold). April 18, 1977, p. 118 (emerging role of gold). February 6, 1978, p. 34 (gold arrangements by Group of 10). May 22, 1978, pp. 145, 151 (gold sales program of IMF). June 5, 1978, p. 161 (gold sales program of IMF). September 18, 1978, p. 303 (gold sales program of IMF). June 4, 1979, pp. 166-167 (gold futures market). September 1979 (Supplement), p. 15-16 (gold sales program of IMF) May 19, 1980, pp. 145, 157 (gold sales program of IMF).

Johnson, Leland L., "The Theory of Hedging and Speculation in Commodity Futures", *Review of Economic Studies,* 27, No. 3, pp. 239-251. Reprinted in A.E. Peck, ed., *Selected Writings,* op. cit., pp. 209-235.

*New York Times.*January 28, 1980, "Gold's New Turbulent Role", Ann Crittendon, p. D-1. September 6, 1981, "Should We (and Could We) Return to the Gold Standard?", Henry C. Wallich and Lewis G. Lehrman, p. 4-E. September 9, 1981, "Gold Won't Pan Out", Robert M. Dunn, Jr., p. A-31. September 18, 1981, "Notion of Reviving Gold Standard Debated Seriously in Washington", Robert D. Hershey, Jr., p. A-1.

Richardson, J. David, *Understanding International Economics: Theory and Practice,* Little Brown, 1980, pp. 54-58, 67-76 and 202-211.

Survey of Current Business, 61 (6) 1, June, 1981, pp. 8 and 50-51

Triffin, R., "The Myth and Realities of the So-Called Gold Standard", *The Evolution of the International Monetary System: Historical Reappraisal and Future Perspectives,* Princeton University Press, 1964, pp. 2-20. Reprinted in Cooper, Richard N., *International Finance*, Penguin, 1969, pp. 38-61.

CHAPTER 4

THE EVOLUTION OF THE GERMAN BANKS' ROLE IN THE FIELD OF GOLD

FRITZ PLASS

1. Relevance of an Analysis on German Banks

To deliver a speech on the role of the German banks in precious metal business is a much more attractive proposition than would at first meet the eye. The first question that sprang to mind was why choose this topic in the first place? Why the role of the German banks? Do they play a particularly special or important role? How could I reconcile that with my personal view that these banks have always been market participants like any others? I found it no easy task to detach myself from this view and reformulate the necessary questions. I intend and indeed I *can* only speak about the path adopted by our bank, Deutsche Bank, because in our country the banks went in for this line of business at different times and in different ways. Why then German banks, why not Belgian or American ones? Certain prerequisites were necessary before a role could be played in a worldwide market such as the gold market.

2. Motives and Prerequisites to Engage in the Gold Market

First and foremost there had to be a freely convertible currency and legislation permitting possession of and trading in gold. Nevertheless these conditions have also been fulfilled in other countries and then without the tax impediments which have restricted German domestic business with private customers.

So other motives and prerequisites were necessary to justify the entrepreneurial decision to engage in the gold market.

Of outstanding importance here is the principle of the universal bank. This principle, which foresees no distinction between, for example, a banque de dépôts and a banque d'affaires, gave banks in the Federal Republic their legal framework. Now these are all necessary technical prerequisites but they are not in themselves enough. So let's also bear the following in mind: you can only build up and maintain a market when there is a corresponding need. And here in the Federal Republic there was truly fertile soil. Twice in the space of a single generation, our country was afflicted by hyperinflation followed by a complete currency reform. Naturally enough this left its mark on people's minds. Tangible assets, not monetary investments, had been the cry. It had not been forgotten. Now it was only a question of channelling this latent need for security along the proper lines. Nothing was better suited to this than gold. With all other investments in material assets, one had to fall back on private sellers or middlemen; the market for the respective goods was complicated and lacked flexibility. So, for the banks to be successful in coupling the confidence people had in them with the right investment media, the new business sector had to be developed quickly and well. As an investment medium gold had a slightly exotic ring to it, which had to be changed. It had to become a quite normal way of investing money, not better but certainly not worse than other investments. Only one type of investment is the correct one in any given situation. Gold was needed if banks were to be able to offer their customers the whole range of possibilities.

3. Domestic Evolution of the Gold Business

But all these considerations I have just mentioned were not put forward as some sort of clever strategy when gold dealing got under way at German banks, they only emerged in the course of the organic growth of this business. The reality of the matter was much more trivial. Permit me now to look back as my topic demands. Consideration was only ever given to what strategy would open up the best market opportunities when the type of gold transaction predominantly practised at that moment seemed to be reaching its natural limits. For this reason permit me to describe how gold business has evolved at our bank and I trust this will give you an insight into the logic behind the path we have adopted.

Round about 1960 practically all the German banks dealt in gold in the same manner. They offered their customers gold coins or ingots as gift ideas and they began to deal in rare coins which they offered to the collectors among their customers. The public at large scarcely paid any heed. Advertising was limited to more or less expensive brochures listing the numismatic coins available and to small display cases with gold coins and ingots which practically only made an appearance at bank counters at Christmas and Easter.

The only real difference between our bank and its competitors was that, for whatever reason, it was often involved in the financing of industrial gold imports for the jewellery industry. But this proved to be the first step into the international gold market. We started offering the gold processing industry not only the financing but also the metal itself. Most of the companies involved were small or medium sized and for them this proved to be a simplification because now, for example, they no longer had to make payments in foreign currencies, or had to fill in import documents. All they had to do was phone the bank and the gold was available immediately. We began to appear as a buyer on the international market. The step into the market for industrial gold was made.

What still remained to be tapped was the latent market for investment gold: private customers. In this context market events in the summer of 1967 showed our bank the path it thought it should take. The outbreak of the Middle East War triggered off strong demand for gold. Not for numismatic coins, but for the simplest and cheapest form of gold. Many banks which had pinned almost all their hopes on numismatic coins, where they could earn higher margins, were not able to satisfy demand. It was at this time that we began to consider whether we should give preference to business in numismatic coins or mass business as an investment medium. We came to the decision that we would only acquire numismatic coins if a customer stated an express wish for them, we would keep few if any in stock and we would limit ourselves to mass business in coins and bullion. But two considerations played a role here. Firstly: investments made at a bank had to have a greater degree of flexibility and fungibility than numismatic coins. The margin between buying and selling prices had to be as narrow as possible and it had to be possible for the customer to be able to realize his investment at any time. Secondly: for a bank with more than one thousand branches, this type of business was much easier to handle. It was not necessary to have specialists in the branches, any cashier was able to conduct this type of transaction.

Since we did not publicize our gold business in any catalogues issued by ourselves, we had to advertise "gold as an investment" and this we did. A large number of other banks who considered that the market for numismatic coins held greater promise became our customers for mass goods and bullion and we became their customers when numismatic coins were required. Now we stopped trading in numismatic items a long time ago but these banks have remained our customers. Once again the signs pointed the way to the international gold market: the best and cheapest way to mass goods was either to acquire them abroad or to have them produced ourselves, e.g. Austrian gold, from bullion purchased on the market. The same was true for small gold ingots. So if the bank was already active as buyer and seller on the international market for gold ingots and mass coins, it was already an integral part of the gold market. The next step to becoming a trading address where we also quoted prices to others was then a small one, simple and logical.

After the introduction of the two-price system for gold which brought the radical chance of strongly fluctuating prices, greater opportunities but also heightened risks, arbitrage business and also the management of the bank's own account gained increasingly in importance. The bank was, after all, a German bank managed at all levels by people sharing the same philosophy towards gold as their customers. If customers were advised to think of gold now and again as an investment, then why shouldn't the bank practise at least to some extent what it preached? Automatically then came in addition to this decision the advising of smaller German banks for their own accounts in precious metals.

4. Relationships with the Biggest Producer: South Africa

The activities that South Africa, the world's largest producer of gold at this time, started to develop were most opportune to our overall concept. South Africa, which had for decades sold its gold solely to monetary authorities, had to develop new markets after the demonetization of gold. A producer, irrespective of what the product may be, must attach importance to selling his product to as wide a public as possible, preferably to buyers who will not sell the product back to the market. Since industrial sales of gold for jewellery etc. could only be increased to a limited extent, the only possibility for expansion that remained was the investment sector. Large-scale investors or speculators would in all probability give occasional consideration to dissolving or diversify their investment in gold and perhaps at the very time the market was least able to absorb it. But if numerous small investors could be persuaded to consider gold as a part of their investment, the danger would be small. The majority of buyers would hold on to their gold as a nest egg. It would be "out of the market". This was the reasoning behind the Krugerrand.

When these coins were offered to established gold dealers they showed little interest. In almost every case they were not branch banks but trading houses, not retail and wholesale traders but only wholesale traders. It didn't fit into their way of thinking, they felt they had neither the time nor the corresponding customers. Now we in the Federal Republic had both. And we were newcomers for neither did the Federal Republic have the glamour of Switzerland as a tax-haven, nor could it rely on a long trading history in gold like London did. Anyway, we had to expand our business if we wished to continue buying in bulk because it's cheaper. But the simple fact of the matter was that both partners, the producer and the distributor, shared the same overall business concept: selling gold to the general public. We assumed sole distribution for the Krugerrand in the Federal Republic. We offered them at attractive prices to all the other German banks and put our feelers out successively in other markets. It all sounds simple but even after we had sold the first million there was still a number of internationally renowned houses which, when asked for a price for Krugerrand, told us:

"There is no market for this coin". But now things have changed. If to begin with we had an almost 75 percent share of total production, this share has steadily dropped although the absolute figures have risen.

5. Moving into the International Market and Becoming an Integral Part of It

That time also saw the very first newspaper advertisement worldwide by a bank for an investment in gold. This was a very important and decisive milestone on the road to the gold market. Gold had made its debut as an investment possibility even for the man on the street. A big branch bank was best suited for this type of transaction. This development resulted almost as a matter of course in larger turnover, heightened interest in gold not only from the general public but also in the bank and in other banks. Now the stage was set, the prerequisites were there, to really be able to play an international role. For the background was there: there was a reliable customer group, we had experience in the field and the risks were calculable.

The step into the international market was supported by a large domestic market. And then there was something that was particularly true for the German banks: the ever-present internal tax disadvantage forcing the banks to be flexible and seek potential ways on all markets. A constraint that others did not have and which is now standing us in good stead in a period of tightening restrictions — we need only think of turnover tax in Switzerland. Once again business was expanded in the three stages we have already seen, however this time no longer in a national but in a substantially larger, international framework: first the gold processing industry, then smaller as well as large-scale investors and last but not least foreign banks and here particularly central banks.

Whereas it had been relatively easy to expand business owing to the relationships that already existed with domestic gold processors, with foreign processors we came up against the well-established competition of international banks and gold dealers.

So here, since everyone had more or less the same purchase sources, we had to offer a service that was different, if possible better, otherwise no one would have switched supplier. An attempt was made with the then new concept of the "gold loan" where instead of selling processors gold and where necessary financing the purchase, we simply provided the necessary gold against interest. This relieved the gold processing industry of all undesired price risks during the production process. We tacitly assumed that the gold needed to cover the loan would be bought from us at a later date. Through these worldwide activities we were compelled to set up a functioning system of settlement, a real freight and insurance department, because, after all, the gold had to be transported as quickly and as cheaply as possible. And if at all possible straight from the place we bought it to the place we would be selling it.

Now if ingots and grain-gold were already being transported for industry and stored abroad, was it not possible to include in the shipment gold coins and small ingots for the local investment market at the foreign location? We did and it brought us additional business. For the third stage dealing with central banks or similar institutions, however, somewhat more was necessary. Here the general philosophy behind all activities on the gold market played a major role.

Firstly we did at no time believe that it is possible to phase gold out of the international currency system.

In a world of inflation, material assets are a good investment. But what other material assets are there for e.g. central banks apart from gold? So here too it was this "broadening of the service offering" that moved German banks to offer gold as an investment to central banks with diversification requirements. As a result of the strength of the D-Mark, the two big German banks conducting gold business became increasingly involved in transactions with central banks. The D-Mark was a favourite investment currency. So it was only logical and consistent that these customers, intitutional investors and central banks, began to settle their transactions via the German banks.

Every bank has developed its own method to ensure that any particular transaction is completed with optimal success for its partner and the bank. We, for example, have always shied away from auctions as far as possible because we always wished to avoid the resultant publicity both for our partner and for ourselves. Far be it from me to say that one model is better than another, it always depends what the principal aim is. Every market participant sets different priorities.

The topic on which I was asked to speak was the evolution of the German banks' role in precious metal business and, wearing a somewhat wry smile, I shall now and herewith have to come back to why this topic was probably selected in the first place: should it have been only the publicity German banks got recently?

Now of course the development had progressed very well but it was pushed very gently and this was a deliberate decision. Bank transactions should always be quiet, quiet and good. But views differ here. While our bank preferred to build up slowly, first on a small scale, nationally, then bigger but still nationally and only then slowly, step by step into big international business, others jumped a few stages. As a result they had to put up with publicity. But, all in all, it took other markets a gratifyingly long time to realize that we, the German banks, had turned into competition.

Anyway the most active big German banks took different approaches to achieve their respective positions in this market. I would not dare to say which approach was the better of the two. Perhaps the market will compel both to adopt a much more similar line in future.

But irrespective of the approach taken, German banks are now an integral part of the international gold market. They operate nationally and internationally, in private customer business as well as with institutional investors. They maintain good contacts with the gold processing industry and are constantly in competition not only as sellers of precious metals but also as buyers vis-à-vis the big producers. The German banks are arbitrage addresses, dealers on the futures markets just like all the others, but as universal branch banks just that little bit different. The future will internationalize their operations even further. Almost all the foreign branches are already involved in this process and, as before, we shall attempt to go our usual way: local industrial business, local investors and banks are to provide the commercial base on which the subsequent arbitrage transactions of the foreign branches can build. Changing circumstances will always compel them to adopt new approaches to achieve their goal.

Only just recently the introduction of the VAT on legal tender coins in Germany has forced us to look for new ways to continue our gold operations in the usual size.

This would perhaps be the best point at which to mention, as an example, the development in Luxembourg. All the big banks and very many other European banks are represented in Luxembourg. For reasons of turnover tax, Luxembourg is used by many market participants now as place of delivery and storage location. It is certainly not yet a real trading centre. To my mind the preconditions are not given for that. The international contacts that London or Zürich or also Frankfurt have are better, there is no national potential such as in the US or also in the Federal Republic given Luxembourg's size and the lack of speculative mentality. So I would tend to say that the significance of Luxembourg as place of delivery and storage will increase steadily, but that the other well-known locations will probably continue to act as turntable for the big trade flows. But that only as a short aside.

As we have seen, the German banks have contributed a good deal towards the popularization and widening of the international gold market. They have found their place in the community of the big precious metal dealers. I hope that they will in future as well be able to use their potential to make the market more stable, less complicated and more predictable. In so doing they will further consolidate their role in these markets as buyers, sellers and administrators of gold.

CHAPTER 5

THE MIDDLE EAST AND SOUTH AMERICAN MARKETS

TIMOTHY GREEN

1. Middle East Markets

1.1 *Changes in the Markets*

The gold markets of the Middle East changed beyond all recognition during the last decade as a result of the immense injection of new purchasing power to Saudi Arabia and the Gulf states after 1973, and the rapid development of communications which has enabled dealers in those countries to trade gold instantly with London, Zurich, New York or Hong Kong. The civil war in the Lebanon also had a profound effect in changing the actual pattern of physical gold flows.

Until the early 1970's Beirut was the prime centre for distributing physical gold throughout much of the Middle East and to Turkey, and was also the channel through which investment or speculative purchases were directed, while Dubai was a market of great importance as the entrepôt for gold destined for India and Pakistan; in its heydey Dubai handled up to 250 tons of gold a year — 25 percent of South African production then. The initial source of this gold was Zurich and London; no Middle East market has enjoyed direct supplies from gold producers, not least because the kilo bars, ten tola bars (3.75 ounces) and smolten lingots that circulate throughout the region are not marketed by South Africa or the Soviet Union.

The picture today is very different. The tragic civil war in the Lebanon ended Beirut's central position as a supermarket for the Middle East. And although some physical gold still goes that way, and considerable gold trading is funnelled through one bank there, it is, sadly, a shadow of its former self.

In its place several regional gold centres have grown up, each offering both physical gold (imported nowadays almost exclusively from Switzerland) and facilities for trading on the international level. Amman,

Jeddah, Riyadh, Kuwait, Bahrain and Dubai are all linked to the international network.

The gold business, both physical and trading, incidentally is very much in the hands of exchange dealers throughout the Middle East, with a few notable exceptions such as National Commercial Bank-Saudi Arabia, Bank Saradar in Beirut and the local offices of the Swiss banks in Bahrain.

1.2 *Physical Trading*

Essentially there are three main streams of Middle East gold activity: the physical game; private investment and speculation on international markets; central bank and government investment institution purchases. Let's look briefly at each.

First, there is the physical business — the grass roots buying — meeting the requirements of ordinary people who still keep much of their savings in high carat low mark-up jewellery (most of which is locally made in workshops) or in small gold lingots. It is essential to realize that these gold purchases are deeply embedded in the social fabric of countries everywhere from Morocco, along the North African coast to Egypt, Saudi Arabia, the Gulf and even Turkey. The farmer who has a good harvest invests the profits in jewellery — I call it "investment jewellery" — and the lady getting married is also judged by the quantity of gold ornaments with which she is endowed. It is not uncommon for a bride to receive ornaments weighing one kilo or more at her marriage — compare that with three grams for a European wedding ring.

This social demand for gold means that the Middle East has long absorbed exceptionally large amounts of gold, considering its relative population. The oil wealth since 1973 has increased the potential and a prime function of regional markets — distribution centres really — such as Kuwait, Bahrain, Dubai, and Jeddah is to supply these needs.

In the late 1970's we estimated that up to 350 tons of gold — effectively 50 percent of South African production — was being distributed by these regional markets. This changed abruptly in 1980 when the price rose to $850, and everyone started selling their ornaments at a profit. In 1980 over 150 tons of gold — scrap — came back to London and Zurich from the Middle East. But in 1981 and today we see again the strong physical demand. We estimate that close to 190 tons was shipped through Beirut, Jeddah, Kuwait, Bahrain and Dubai in 1981, quite apart from up to 50 tons of jewellery from Italy: perhaps 240 tons in all — well over one-third of South African production. And more recently purchases have been at an even higher rate: we estimate that at least 25-30 tons of gold moved to the Middle East for the "grass-roots" demand in January 1982 — an annual rate equivalent to perhaps 300 tons. That is one measure of the Middle East's importance in gold: it is a cornerstone of the physical demand.

1.3 *The Investment-Speculation Activity*

Secondly, the Middle East in the last decade has become a major source of investment/speculation activity in gold on the London, Zurich, New York and Hong Kong markets. This demand has been channelled primarily through a handful of leading banks and exchange dealers in Beirut, Amman, Jeddah, Kuwait and Bahrain, who have become major traders in the international market place. The financial resources behind many of these traders, especially from Saudi Arabia and the Gulf, have given them tremendous muscle. And in 1979 and 1980, in particular, they were often involved in substantial speculative forays which gave great impetus to the price.

Indeed, it is important to realize that it is the driving force of this Middle East speculative initiative which influences the price, rather than the day to day physical demand, which at best, can help only to hold or stabilize the price. The significance of the Middle East's role in trading was underlined to me by a major New York bullion house, which reckoned 70 percent of the business it was placing on Comex in 1980 came from Middle East clients.

The growing enthusiasm for gold trading during the late 1970's not only encouraged exchange dealers to install sophisticated trading rooms, which often became social centres for local speculators eager to play on Comex in the evenings, but also brought much business to American houses which opened offices in the Gulf.

Those forays, however, were not always so successful. A great many speculators lost a great deal of money. And the picture today is rather different. Middle East investment or speculation in gold early in 1982 is at a very low level. As a banker in Bahrain told me in January, "Investment interest in gold just now is zero". The action, for the moment, is in local stock markets. But that lull will not last forever, and when the next bull market in gold develops you should expect to see Middle East money back in the game.

1.4 *Central Banks and Government Demand*

Finally, and briefly, we must not forget the role of Middle East central banks and government investment institutions in the gold market. Several of them have decided in recent years to keep a small part of their reserves or portfolios in gold — 7 percent to 12 percent seems the fashionable amount. And in each year since 1979 they have been net buyers of modest amounts of gold. Such transactions, however, are undertaken on the international markets and not in the regional centres. But they must not be ignored in assessing the role of the Middle East in today's gold market.

And a key question over the next two or three years as oil surpluses tend to diminish is: "Will the banks still buy gold?" The room to maneuvre may be less. For the moment, at the beginning of 1982, it is this modest official demand together with a high level of physical offtake through the regional distribution centres that provides the Middle East's main influence on gold.

2. Latin American Markets

The markets of Latin America are somewhat on the sidelines of gold, because most countries there are self-sufficient in local gold production, and thus do not need to rely on international markets.

Moreover, apart from some special exemption, they have not yet become substantial traders in Europe, Hong Kong or New York. Relatively little money from this area went into the great gold speculation of 1979-80, compared to the Middle or Far East.

On the strictly physical level, one exception is Mexico, which regularly imports gold from New York and Toronto, both for its local jewellery industry and for its far reaching coin-making programme. The Bank of Mexico used over 40 tons of gold for coin-making in 1981.

Argentina has also been edging into gold during the last three years since controls have been lifted. A modest gold coin business has developed, chiefly in sovereigns and Mexican coins, but recently in Krugerrands, which were 'launched' there late in 1981.

Panama is also a modest gold market, distributing physical gold to surrounding countries, and acting as a channel for some speculative buying from Central American states. Panama is also, incidentally, an important distribution centre for Italian gold jewellery.

Potentially Brazil could become the most interesting centre, if the widely proclaimed gold discoveries there are really fulfilled. Brazilian gold production more than doubled between 1978 and 1980 to 35 tons, and may rise towards 50 or 60 tons during the 1980's, making Brazil the second largest producer in the non-communist world. That is more than the local markets of Rio de Janiero and São Paulo require for jewellery or investment, and it is thus possible that Brazil could become a substantial gold exporter, unless the central bank buys up all the surplus gold as it is attempting to do now.

One drawback so far is that no Brazil refiner has yet won "good delivery" status for his bars, so they do not gain automatic acceptance abroad. But much more modern refining facilities are now planned and undoubtedly Brazil is the country to watch over the next decade.

CHAPTER 6

MARKETS AND INTERMEDIARIES IN THE FAR EAST

DICKRAN GAZMARARIAN

1. Growth of Hong Kong as a Major International Market

I am sure that most of you here today are aware of Hong Kong's standing as a major financial centre in the Far East and also of its well established gold market which is capable of handling all facets of the trade on a regional and international basis. Looking back it is sometimes difficult to grasp quite how rapid its development has been. Seven years ago, when Mocatta first opened its office in Hong Kong, it was difficult to find more than a handful of counterparts with whom one could deal. Today, we have reached the stage where almost every major bank and bullion dealer has a presence in the Colony, all playing their part in the continuing growth of our market, and those who are not represented, cannot afford to ignore its existence and avail themselves of its services from all corners of the world. Though Hong Kong's rise was dramatic, it was also natural that Hong Kong, rather than any other country in the Far East, would develop as the region's foremost gold dealing centre as Hong Kong has traditionally been the focal point of the trade. The Chinese, like other races and nationalities, have a historical affinity to gold and silver and Hong Kong has serviced the trade for the past 80 years. The decision by the Hong Kong government in 1974 to lift all restrictions on the import and export of gold set the scene for Hong Kong's rapid and unrestricted development into a truly international market. Hong Kong was able to capitalize on the liberalization almost instantly as it possessed the financial infrastructure necessary to support its development, and more importantly an organized and successful gold market was already in existence in the form of the Chinese Gold and Silver Exchange Society.

2. The Chinese Gold and Silver Exchange Society

This exchange proved to be the cornerstone on which the market broadened and developed into what it is today and the arrival of the international bullion dealers tied this market into the other world gold centres.

The Chinese Gold and Silver Exchange Society was established over 70 years ago, and until the liberalization in 1974, it principally catered for demands of the regional gold trade. It was the only exchange open during the Far East time zone and was therefore used extensively by traders, speculators and smugglers alike, who all needed a market to satisfy their various requirements.

The Chinese Gold and Silver Exchange Society, which has 195 members, is an exclusive Chinese preserve. Dealing is conducted by open outcry in Cantonese with prices quoted in Hong Kong dollars per tael of 99 percent fineness (the tael being equivalent to 1.1913 oz. of pure gold) and transactions are for 100 tael units. To an international dealer the difference of weight, fineness and currency presents no problems, but thereafter the mechanism of the market is quite different and unique when compared to western markets. It is essentially a spot market and an undated futures market at the same time. The price one deals at is for spot gold for same day settlement, but there is no obligation on the part of the long or the short to make or take delivery. The rules allow all open positions to be automatically rolled forward for one day at a time, for as long as one wants to keep a position open. Naturally there is a mechanism to take into account the "carrying cost"as we know it on futures markets, and this again takes place at a daily "interest fixing" session. During the fixing longs or shorts may also declare their intention to either deliver or take material but with no obligation for the other party to do so. If, for example, a short wanted to deliver 10,000 taels but no longs wished to take it, then the long would be asked to pay *x* number of Hong Kong dollars per tael for refusing to do so, and the short would receive the equivalent amount. In times of high demand the reverse could happen (a backwardation) whereby the long would want delivery and the shorts were unable to meet the demand, then the shorts would be penalised in the same fashion by being charged *x* number of Hong Kong dollars per tael per day for as long as they were unable to satisfy demand. The "interest fixing", now that Hong Kong has become an international market place, normally follows the carrying costs of other world markets and is closely related to the Hong Kong dollar funding costs, much the same way as Euro-dollar rates are reflected in the forward Loco London quotes. However, just last month, when gold dropped to around US $370, there was such heavy physical demand for taels throughout South East Asia, that our market went into a backwardation whereby the longs received 11 percent per annum whilst the shorts were unable to satisfy demand.

3. The Development of the Loco London Market

The Exchange has operated in this fashion, without a clearing house system, extremely well and at the height of the gold market in 1979-80 it was not uncommon to have turnover in excess of a million taels a day. The exchange has traditionally drawn business from all parts of South East Asia which accounts for the depth and liquidity of the market. It is not surprising therefore, that when gold was completely liberalized, international bullion dealers soon turned their attention to Hong Kong, particularly to the Exchange, and saw the possibilities it afforded. The attraction, given a well established exchange and its supporting financial services, was that Hong Kong was in an ideal time zone, 8 hours ahead of GMT, for dealers to make a market for gold in a currency other than in Hong Kong dollars. The primary function of the first bullion companies that established offices in Hong Kong from 1976 onwards was to create a dealer market for gold quoted in US dollars per ounce, which in effect became an extension of the Loco London market in the Far East. The Loco London market was able to develop rapidly mainly due to the fact that the tael market was very much a market of "last resort". In the early days of the dealer market in Loco London all deals concluded with regional or international counterparts were invariably offset on the tael market which was liquid enough to absorb large transactions. The ability of the Hong Kong dealers in making narrow and competitive quotes encouraged more and more European, and later American, traders to use our market and led to its international acceptance and recognition. The Chinese tael market also provided the international bullion houses with interesting and lucrative arbitrage opportunities. The different standards in fineness of gold, currency, and location were all the ingredients an arbitrageur needed. However, due to the rather unique nature of the tael market and its interest fixing session it was impractical to arbitrage taels versus other markets from a distance. Market conditions change so rapidly, from day to day, that it is essential to have a presence in Hong Kong if one is to arbitrage and use the tael market successfully. The Loco London and tael market complemented each other ideally and the relationship between the international dealers and the local member brokers of the Exchange was and still is today very harmonious. As more and more banks and bullion dealers established themselves in Hong Kong, all essentially making a market in Loco London, the reliance on the Chinese Exchange as a market of "last resort" naturally diminished. However, on the other side of the coin, a considerable number of Hong Kong-based Chinese companies themselves soon became market makers in Loco London in addition to acting as brokers on the Exchange. This further enlarged our market and Loco London trading spread from Hong Kong into other South East Asian countries. I think it would be fair to say that we have reached the stage today where both markets are on equal terms as far as volume and influence are concerned. The market has become extremely competitive but we have a happy inter-relationship between local, regional and international participants.

4. The Development of Secondary and Third Markets

As Loco London trading became a familiar trading medium in Hong Kong and other South East Asian countries, and as public interest in the metal grew considerably in the late 1970's, it soon became evident that Loco London trading, both for spot and on a deferred basis, had one important advantage over other markets — that was that one could deal in the same "contract", as it were, for 18 hours a day and this of course became possible with the emergence of Hong Kong and New York as major dealing centres. The erratic markets in 1979 and 1980 made this an even more important factor in the minds of traders and speculators anywhere in the world who had to monitor price changes almost around the clock.

This has been borne out by the development of the "secondary market". Commodity brokers throughout the world soon found that there was a growing demand from their customers, traditionally futures market traders, to provide facilities for them to trade gold on a worldwide basis, rather than merely acting as agents for futures business. The brokers found that the best medium to satisfy this demand was in fact Loco London gold which could be traded round the clock through their international network of offices. I call it the "secondary market" in that it is once removed from the wholesale or primary market of the traditional market making function of the major bullion dealers.

Gradually, both local and international brokers secured lines and facilities for deferred spot trading from the major bullion dealers and they in turn passed on these facilities to their own customers as brokers, by nature of their business, are best suited in handling the retail side of the market. In Hong Kong today, almost all the large international brokers, and a growing number of local brokers, offer their clients a Loco London gold trading account, either for spot or a deferred settlement basis. The relationship between the brokers and bullion dealers is naturally very close, and retail business generated by the brokers will eventually be covered with the dealers. This development has added a new dimension to the gold scenario which has increased the scope and depth of our market even further.

Brokers today not only offer their services for traditional futures business but have also become market makers in a secondary capacity to their own retail customers. Today, we have in Hong Kong over 15 major bullion dealers, 7 international brokerage firms and over 40 locally based brokers all actively engaged in the gold trade. These companies, coupled with the members of the Chinese Gold and Silver Exchange Society, draw business from Japan, Australia, Singapore, Taiwan, Indonesia and Thailand, as well as from Europe and North America, and each services the needs of a particular segment of the industry. Jewellers in Hong Kong and South East Asia turn to the tael market for their needs, a trader in New York uses the Loco London market to cover an overnight exposure, a speculator turns to his broker to cover or initiate a new position. The combination and interrelation of all these participants has made Hong Kong what it is today, a large and viable market place.

We do have a third market which I would like to briefly touch upon. In 1976, the Hong Kong Commodity Exchange was established and initially started by trading in cotton and sugar futures. In 1979, a gold contract was added which was modelled on the lines of the New York Comex market. Prices are quoted in US dollars per ounce of 995 fine gold and trading is effected in lots of 100 ounces for delivery in Hong Kong. The traded months are the same as Comex and trading is by open outcry in Cantonese. The market has not been a great success so far with turnover currently running at a modest 100 lots per day, which is insignificant compared to the tael and Loco London markets, but the Exchange is well structured and I believe that it may still become a significant market in the next two to three years.

5. Hong Kong's General Financial Structure: a Fundamental Framework

I said earlier that the predominant factor that established Hong Kong as the major gold market in the Far East was the existence of the Chinese Gold and Silver Exchange Society, but the reason it was able to blossom so rapidly was in no small part due to Hong Kong's well established financial structure. Excellent telex and telephone communications made Hong Kong readily accessible to the rest of the world. The low corporate tax structure was an incentive for companies in establishing offices, and the lack of exchange controls permitted the arbitrageur or trader to deal in any currency, be it in yen, US dollars or Hong Kong dollars.

Local Chinese banks have had a long and close association with the Chinese Gold and Silver Exchange Society and have played an important role in its efficient operation. They are the main depositories for gold and their understanding of the complex rules of a rather unique market encouraged them to finance the traders on the Exchange in either metal or money. You will recall that the tael market is traded in Hong Kong dollars and the daily "interest fixing", which sets the carrying cost for gold, is closely related to overnight money rates, much the same way as Loco London forwards are related to the Euro-dollar rates. To prevent undue distortions in the "interest fixing" local banks will lend gold or money into the market. Naturally rates do go out of line from time to time, but ultimately the banks' involvement will re-establish the market equilibrium.

When international bullion dealers arrived in Hong Kong, they too needed the services of the banks for funding their arbitrage activities and in providing foreign exchange facilities.

In more recent years, a growing number of banks in Hong Kong have started retailing gold over the counter to cater for the growing investment demand from the general public for gold in smaller, more affordable units. Typically the mix is from Krugerrands, Maple Leafs, 5-50 gram bars, gold certificates and Passbook schemes either in ounces or taels, and today over 15 banks in Hong Kong are actively promoting this side of the trade. Traditionally the Chinese public have had a preference for the small 1 tael - 5 tael bars which can be bought from most jewellery shops, but since Intergold established their regional headquarters in Hong Kong in 1979 and actively promoted the Krugerrand, the coin has become increasingly popular with the Chinese.

6. Other Far East Centres: Japan, Singapore, Australia

As Hong Kong was growing into an international market place, so too was the trade developing in Australia, Japan and Singapore as foreign exchange restrictions and trade controls were gradually eased or abolished.

Australia and Singapore both have well established futures markets modelled on the lines of New York Comex, but which by comparison are not as active. Singapore has developed an active inter-dealer market in Loco London which complements its physical entreport trade with neighbouring countries, and its success has been due to the active participation of most of the major Singapore banks. Japan's role and importance in the gold markets grew significantly in 1981 and the authorities' policy of gradual liberalization of the trade is set to continue in 1982. Under the supervision of the Ministry of International Trade and Industry, Japan will establish a gold futures market by the end of April. At the same time, it is anticipated that Japanese banks, that have hitherto been prohibited from dealing in gold, will be allowed to do so when the banking laws are revised by the Ministry of Finance. The banks' initial participation will be limited to the retail over the counter sales of kilo bars and smaller gram bars.

All these developments will continue to enhance the importance of the Far East as a major gold trading area with Hong Kong remaining the focal point and price leader of the region.

CHAPTER 7

THE OPPORTUNITIES AND RESTRICTIONS FACING NEW INTERMEDIARIES IN THE INTERNATIONAL GOLD MARKET

ROBERT GUY

1. The Problem

You have heard this morning five geographic viewpoints on the recent developments within the international gold market. It is my task to draw general conclusions from these papers which may be of some assistance to those institutions which are contemplating participation in the market as new intermediaries. By its very nature it may be a somewhat presumptuous task as I have not had the benefit of reading their papers before writing my own speech. Furthermore the very diversity of the market, of which you have been made aware, precludes the preparation of a definitive blue-print for the establishement of new market-places. "One world" yes. But it would indeed be a dull one if a quest for uniformity ignored the potential for local imagination, innovation: the inevitable difference in local character and customs. There is always a need to learn and indeed benefit from the experience of others.

What then may be learned from the experience of the international gold market in the last decade? What proven historic guidelines have been of benefit to dealers in gold and what new lessons have been learned in recent years? What should new intermediaries expect of their own central banks, government departments or agencies and will self-regulation remain the prime necessity or will the very entry of an increasing number of new participants lead to increased legislation?

2. Lessons from the Last Decade

Three questions, so let me quickly, in answer to the first one, review the experience of the last decade. Was the expansion of the market inspired or inevitable, planned or undisciplined? In my view what we have seen is a natural and indeed inevitable commercial response to the changes in the international monetary system — the reappraisal of gold as a reserve asset — and, during that period, to the evidence of governmental inability to come to terms with widespread inflation. Such a response may well have been, in part, instinctive, but this does not mean it was either irresponsible or undisciplined. Trading in gold, as you have heard from the other speakers, is no recent phenomenon; the skills are long established. The increased capacity to demonstrate such skills was, of course, frequently enhanced by changes in local legislation; the gradual easing of relevant exchange control laws in Japan beginning in 1973, the legalization of international gold trading in Hong Kong in 1974, the liberalization of gold in the United States in 1975.

Such changes in domestic legislation coincided with the international intention to demonetise gold and the official exit of the central banks from the market-place created an additional opportunity for the commercial banks. As it happened this opportunity proved greater than most expected as we entered a period of great turbulence in the exchange markets which culminated in fears of excessive US inflation to which many — both official and private — investors responded by purchasing what they considered to be the ultimate store of value — gold. Now that we are in a period of declining gold prices activity of the market has decreased but this has not been paralleled by a decline in the number of major participants; indeed the demonstrated strength of the market through this turbulent period continues to attract new intermediaries.

3. The Strengths of Existing Participants

Turning to the second question, therefore, what have been the strengths of existing participants both in terms of historic guidelines and in their ability to react to new circumstances?

The crucial point for new intermediaries — particularly banking institutions which normally deal in paper currencies — is to remember the importance of the *physical* market. A premium has to be placed on stock management; storage, refining capacity, transportation, availability not only in location but also in the right form. Many of you will be aware in this latter context of the increasing number of refiners whose name has been added to the London Good Delivery List in recent years thereby facilitating the international trade of the so-called standard (400 ounce/12.5 kilogrammes) bars. It should be noted however that this is not a status easily achieved and indeed our London Brochure has recently been revised particularly to underline the stringent requirements.

The preparation of the London Goods Delivery List is but one example of the way in which the London gold market fulfills a greater role than that of a broker. We also effectively provide the international clearing centre for the market; the international price is the "Loco London" price and settlement is effected across the exchange accounts which all significant intermediaries need to maintain with members of the London market.

Apart from the understanding of the physical market, gold dealers must follow the criteria well established in the parallel markets of foreign exchange and currency dealing. After a period of intensive activity in interest-arbitrage it is interesting to note the reminder the market has received during the last year of the need for skills in its market-making and client liaison. Dealing expertise has, of course, to be accompanied by clear management guidelines both in terms of credit exposure and of dealing risk. It is, I suggest, not for major financial commercial banking institutions to maintain high speculative positions to the potential detriment of their shareholders, the market as a whole or indeed to those who work for them on the dealing desks. As for the dealers themselves, I personally believe that they should not be active traders on their own account. This, of course, was the view expressed in London by the Governor of the Bank of England in 1980 and the maxim "liberty without licence" still stands.

Such truisms — which does not make them clichés — have been, as I say, part of the historic guidelines followed by gold dealers. More recently answers have had to be found to new questions posed by an evolving market-place. Such answers are not always easy to find as has been experienced by those of my colleagues in London responsible for the creation of the London Gold Futures market which it is now anticipated will make a successful opening within the next few months. For the purposes of this paper and given the restrictions of time it is perhaps of interest to look at one particular conclusion which has been reached — the desirability of an independent as opposed to a mutual guarantee. The creation of a completely separate clearing house does, we believe, enhance its objectivity and credibility. By their very nature, the principal participants in a futures market hold different perspectives; some are motivated by immediate commission earning opportunities, some by the exploitation of medium term arbitrage strategies; some enjoy the support of substantial capital, others are more limited — although there are, of course, stringent rules on minimum capital requirements; some, by tradition, are primarily market-makers, others more usually fulfill the role of brokers. Such a diversity of backgrounds could at times lead to an internal conflict of interests between the members themselves which in turn could lead to a further external conflict between the market itself and those who utilise its services.

LGFM has also attempted to overcome what some consider the disrupture effect of "Limit" movements. When these occur, the market will close not for the rest of the day, but only for an interval of half an hour.

The confidence in the viability of a European futures market is greater in

London than Zurich. We believe that (i) there is a time-gap to be filled, (ii) the institution of such a market will give a new dimension and depth to the European market as a whole and (iii) there will be a natural synergy with other emerging futures markets such a LIFFE. We do, of course, speak against the background of active experience of the gold and silver contracts on Comex over many years and indeed we are all (either directly or indirectly through subsidiary companies) members of that particular US exchange.

This practical experience makes us mindful of the need to counteract the accusations that futures markets may inspire excessive volatility. Some of course claim the contrary and say that such markets provide a smoothing function. The debate is still open. In theory the lacking of forward cover should temper price movements but the performance of the Comex silver contract in 1979-80 would appear to contradict this view as would the activity, from time to time, of some of the foreign exchange contracts on the IMM Chicago.

4. Self-Regulation and Official Guidance

A discussion of these points leads naturally to the third question — what may new intermediaries anticipate on the part of central banks, government departments or agencies, given of course the natural supervisory role of the market-place which will be delegated to them? I would anticipate that the general policy of self-regulation will continue. Self-regulation is a term that can easily be misunderstood because some — wrongly — believe it entails total freedom from official guidance. Others, noting the increasingly legalistic tendencies of the commercial world, consider that it must be somewhat anachronistic; should we, they ask, really still be relying on guidance by way of nods and winks emanating from the window if you will excuse this mixture of Japanese and English metaphors. The dilemma facing government institutions is how to ensure the continued efficiency of market-places which in the event of failure would rebound to their own discredit and might necessitate their own financial support whilst at the same time recognising that they themselves do not have the depth of practical experience required to operate such markets?

The answer appears to me to lie in self-regulated markets maintaining regular liaison with the relevant authorities, seeking views from them and bringing to their notice problems of which they may not have been aware. There is a need to draft rules and regulations which are not inimical to their interests, to take account of their own advice on financial limitations and to impose a code of conduct which draws heavily upon the experience of parallel markets.

Some wonder if it is necessary for established futures markets to create a compensation fund to protect the small investor. The small investor however tends to be primarily vulnerable to the fraudulent practices of so-called broking firms which are not members of those markets through which they themselves operate. He is better protected by appropriate company legislation rather than a compensation fund provided by the market. A properly established futures market rests its case for credibility on the strict financial requirements for membership and the need for members to become, in turn, clearing members of a nominated clearing house. Furthermore the essence of a futures market's capacity to perform lies in the requirement for mandatory margins to be segregated and deposited with the clearing house.

The main point, however, as I say, is to maintain contact on all such subjects with the relevant authorities. It is interesting to note that the problems in the US futures markets gave rise to greater interventionism on the part of the CFTC, but with both sides, government and market, now more aware of each other's standpoints, the essential spirit of self-regulation has been maintained.

Such dialogues ensure that the development of futures markets will continue, that they will be self-regulating, but that central banks, government departments or agencies will maintain an important supervisory role. It is also possible that potential intermediaries in the gold market will see on the part of major central banks not merely a concern to supervise the market but also to participate within it.

5. Central Bank Intervention

We have, of course, over the last six years or so, seen significant central bank activity but this was in the main conducted as part of an overall reserve policy — buying in order to diversify assets or selling (Canada being the most notable example) in order to rectify what was considered to be an over preponderance in gold. The case for entering the gold market in order to assist stability may now be considered stronger. It is a case which has been presented by some commercial bankers for some years, was propagated by Dr. Leutwiler as President of the Swiss National Bank in 1979 and was supported by Dr. Zijlstra, President of the Netherlands Bank, in 1981. Whether such action should be concerted (by the revival of a Gold Pool) or undertaken on an individual basis is still open to debate. The workings of the EMS show the benefit of concerted action but gold, as it were, is a currency of wider circulation and its movements in value of importance to a greater number of countries.

Whether discussing paper currencies or gold, it is surely accepted by an increasing number of observers that one should not approach the arguments over exchange rate management as, to use the current English political parlance, either a "wet" or a "dry"; there is an alternative between fixed and floating rates. In the exchange market, totally free floating can, for example, lead to an over-valued rate causing severe commercial disruption which is permanent and not abated by a subsequent market adjustment. Conversely excessive devaluation leads to the likelihood of imported inflation which may not be controlled for a significant period of time. Excessive volatility is also dangerous to the markets themselves and to those which participate within them as a service to their customers. Danger leads to illiquidity which is then compounded by fear and irrational expectations. Intervention can temper such expectations and restore some stability. It should not be undertaken to impose new values for either currencies or gold because values are relative and shift constantly. The expectation of changes in value are, of course, the attraction of any market place. Indeed one might even say that they provide some "magic"; but "magic" is one thing, dangerous hysteria quite another.

6. New Intermediaries

Seven years ago, while looking into the future, I forecast the eventual participation of major US commercial banks in the gold market (there are now three) and speculated, if that is the word, on the likely creation of a gold futures market in London. It is always best to remember only the best forecasts or, alternatively, to make so many that some are inevitably proved correct. On this occasion it would seem appropriate to conclude that in my personal view the market will continue to attract new intermediaries, that in general they will operate in a self-regulatory environment and that whilst any market-place may be dangerous excessive movements in the price of gold may be tempered by constructive central bank intervention. At least on the entrance to this market, there is no need to issue here in Rome and out of respect to our Italian hosts BNL and Nomisma (to whom we are so grateful for organizing this Conference), Dante's warning "Lasciate ogni speranza voi ch'entrate..." All hope abandon ye who enter here (Divine Comedy: Inferno). Provided they follow guidelines such as those which have been itemized today, new intermediaries can approach the gold market with confidence.

CHAPTER 8

MARKETS AND INTERMEDIARIES: CONCLUSIONS AND EVALUATIONS[1]

LAMBERTO DINI

1. Gold Markets as an Expanding Reality

It is my duty to report to you on the first session of the Conference which was devoted to the analysis of gold markets and of intermediaries that deal in gold internationally.

I think that the first conclusion that can be drawn from those debates is that the gold market is indeed an expanding reality from a geographical point of view.

Alongside the traditional markets in London and Zurich, new centres have acquired international prominance in gold trading: I would mention first of all the United States market — which from 1975 has begun to play an increasingly important role, particularly in the field of gold futures — but I would also draw attention to other emerging markets. Dickran Gazmararian unveiled for us the special features of the Far East markets and the rapid changes during the last decade which have made Hong Kong one of the largest gold markets in the world. And we have also heard from Timothy Green about the growing impact on the gold markets exerted by the capital surplus Gulf states, and about the great potential, as an international gold centre, of a country like Brazil.

Thus, we can say that in recent years gold has acquired a broader international status and that gold markets are now much more articulated as an institutional framework than they were only a few decades ago.

[1] The conclusions and the evaluations of the Session on Markets and Intermediaries of the World Conference on Gold were drawn, at the end of the Conference, by Lamberto Dini, Chairman of the Session, while a final comment was made by Guido R. Hanselmann, opening speaker of the Session.

2. The Structure and Behaviour of the Market

The second major conclusion that can be drawn regards the structural features of the two segments of the gold market, which have acquired clear-cut characteristics over the recent past. I refer specifically to the spot market for physical gold and to the futures market. As to the behaviour of these markets I believe there is a large consensus among our speakers on their main differences and interrelations. The spot market appears to be characterized on the supply side by a monopolistic situation, with South Africa and the Soviet Union playing by far the largest role. On the demand side we have a variety of motivations represented by the demand for industrial use, for jewellery and hoarding. As Dr. Hanselmann clearly explained in his very lucid presentation, the main factors affecting price conditions in the spot market are the long term trends in demand and supply.

The second segment of the market, that dealing in futures, is a relatively new development, and in a geographical sense is presently concentrated in the United States. A number of speakers have explained the various roles performed by the futures market and have answered some of the accusations which are voiced against this market. I think a consensus has emerged as to the effect that the futures market does provide a useful vehicle for investors to diversify their portfolios by entering into highly liquid and potentially profitable placements. In this connection it was also pointed out — notably by Raymond Nessim — that the futures markets are in fact a system to capture liquid funds which will be eventually rechannelled into productive uses, thus responding to the accusation that the futures market subtracts financial resources from productive investment.

3. Futures Markets: Stabilization or Destabilization?

More generally, an interesting point of our debate was whether the futures markets are — so to speak — able to take care of themselves and whether speculators in futures stabilize or destabilize the market. While it is recognized that, because of their enormous size, futures markets have influenced in fact price developments at least in the short run, most speakers denied that futures markets have a destabilizing effect or tend to produce excessive and erratic price movements. It has on the other hand been maintained that speculation does contribute to smooth the wave of price fluctuations since, if I may use a happy turn of phrase by Dr. Hanselmann, "Speculation has the function of taking over the risk from consumers and producers".

I think we should all agree on this interpretation and on the fact that futures markets have generally been able to regulate themselves rather successfully.

4. On the Supervisory Role of National Monetary Authorities

A rather explicit call for a more active supervisory role by national monetary authorities has been made by Robert Guy. I would like for a moment to make a few personal remarks on this topic, not so much as Chairman of a session of our Conference but as a central banker. I agree that the outlook for the futures market points to the possibility of a substantial expansion both in volume and geographical distribution. I believe that London is likely to play an increasingly larger role in this area, and here I thus tend to agree with the prediction made by Robert Guy.

Such an expansion of the futures market should be seen as a positive development, although it inevitably entails significant consequences for the monetary and exchange rate policies of individual countries. This said, however, I would not go as far as Robert Guy in advocating central bank interventions in the gold markets, be they spot or futures. In my view several reasons militate against such an action. Even leaving aside the international agreements forbidding central banks to act in such a way as to manipulate the price of gold, I think it would be unwise for central banks to intervene in markets on which — contrary to what happens in foreign exchange markets — they have no control whatsoever on the underlying determinants. I am afraid that such "intervention in the dark" would be more destabilizing than helpful; also I would have problems in identifying what could be the possible guidelines for such intervention policies (I am aware that views differ on this point and we have heard some pertinent and stimulating suggestions by Professor Quadrio-Curzio this morning). This of course does not mean that gold markets should not be supervised in any way. But supervision should be clearly confined to the establishment of prudential standards and to the provision of statistical information on the size of the market and the terms and conditions of its operations. These, I believe, were the major themes that emerged from our discussions.

5. On the Issue of a Return to the Gold Standard

Other specific issues were also raised, which were perhaps within the competence of other sessions of the Conference and which would be certainly dealt with by their respective Chairmen. I refer in particular to the role of gold as a reserve asset, to the question of the gold content of the ECU, and to the future of the SDR.

I would like just to say a few words about the issue of a return to the gold standard to which reference has been made by several speakers during the first session. There is no doubt that, as indicated by Fritz Plass, it is impossible to phase out gold from the international financial system. For the reasons that were indicated during our discussions, gold and gold-related assets seem to be very well entrenched in our economies, and their role is far from declining. This however does not mean that the trend is for a return to an international gold standard, perhaps quite the contrary!

In my view the fact that a Gold Commission has been established in the United States should not necessarily be interpreted as a move in that direction. In fact I believe that not even the proponents of such a Commission had really in their minds a return to gold in the international sphere at least in the short run. What is behind the gold proposal is in my view a problem of domestic monetary control and the desire to compel monetary authorities to follow sounder monetary policies by subjecting them to strict rules. While I do share the aims of such proposals, I also tend to share the doubts expressed by Raymond Nessim as to whether such a system would really stabilize money supply.

Be that as it may, the rapid development of private gold markets and the sheer size of these markets would seem per se to have placed a permanent spanner in the wheels of the gold standard advocates.

GUIDO R. HANSELMANN

6. On the Supply Side

Just a few additional short comments on the conclusions that can be drawn from the discussions on markets and intermediaries.

First, on the supply side, I have two observations. The first is the reference to the monopolistic organization of that market, because only the Soviet Union and South Africa are principal suppliers. I made reference to this in my speech and Dr. Dini, in his summing up, again repeated it. I feel it is a qualified statement as far as South Africa is concerned. South Africa has been a regular seller and it does not appear that it does take advantage of this monopolistic situation. So I think that in fairness to our friends in South Africa I should have made this qualification.

The second observation on the supply side is that the higher prices that have been in existence now for a number of years, are clearly boosting production in some countries. Peter Fells[2] has given the figures on the estimated rate of production in some areas, but I think it is significant that, for instance, Canada, which produced 50 tons in 1980, is supposed to produce 84 tons in 1985. The United States is expected to increase its production from 27 tons to 57; Brazil from 35 tons to 75 and Oceania from

---

[2] See Louise du Boulay and Peter D. Fells, *Medium and Long Term Structural Aspects for Gold*, Part II, Chapter 9.

32 tons to 63. I think this is a most remarkable development, highly welcome because it broadens the market.

7. On the Demand Side

Now a few observations on the demand side. I think the main feature — Dr. Dini has already referred to that — is that the markets have expanded very considerably in the course of the last 15 or 20 years. I take the liberty to elaborate on that a little bit. Fifteen years ago we had two principal markets: Zurich offering all related banking services and catering — as Paul Jeanty said — to the retail business, and London which — also according to Paul Jeanty — does the wholesale business.

London will certainly increase in importance again through the introduction of the futures market. I do believe that the main attraction of the futures market in London will be that it will be outside US controls.

Now today, in addition to Zurich and London, we have seen emerging important other centres; the importance of Hong Kong has been described by Gazmararian. As for New York, we have discussed at length the importance it has for the futures market. And although both Jeanty and Oppenheimer[3] referred to the futures market in New York as a "casino", I maintain that it fills a very useful purpose.

Coyne[4] has given some information on Tokyo, a very fast growing new market which in 1981 imported about 5 million ounces. He also pointed to the problem that still exists in Tokyo, namely the infrastructure.

The last place I would like to mention here is Luxembourg, a new trading centre. Plass cast some doubts on the potential in Luxembourg. I do share these doubts because I think that the creation of the Luxembourg market is more or less a bit artificial.

We have not discussed two other very important markets during our meeting. I am referring to Canada and Singapore.

I do believe that we should also attach importance to these two markets. Canada is a principal producer and a consumer and we should not forget that the first futures market and the first options market in the world for gold were created in Winnipeg, Canada.

Singapore evidently has special importance because of its trading functions for neighbouring countries, particularly for Indonesia.

As a result of all these developments over the past 15 years the markets

---

[3] See Peter Oppenheimer, *Possible and Desirable Role for Gold in the International Monetary System: Reconsideration and Proposals*, Part III, Chapter 18.

[4] See Herbert J. Coyne, *World Monetary Demand for Gold: Key to the Gold Price Outlook*, Part II, Chapter 11.

have become a lot wider, they offer a lot more capacity to transact large volume. I think that this, together with the existence of the futures markets, would allow central banks to place business of the kind that Professor Niehans[5] suggested in his address yesterday afternoon.

Trying to establish a matrix comparing the functions of particular markets against the main sectors of demand would probably show the following picture. Industrial demand is principally covered out of Zurich, London and New York. The demand for jewellery manufacturing, which was so interestingly described by Stella and Torboli[6] as regards the Italian market and Timothy Green when he referred to the Middle East markets, of which he said that they had changed beyond recognition in the course of the past years, is principally covered out of Zurich and Hong Kong. For the private investors there is a wide choice of markets: Zurich, London, Hong Kong, New York, Tokyo and Singapore, whereas the central banks, the first large segment of demand, will principally cover their needs through Zurich and London.

8. Markets Growth in the Future

The question for me is which sector and which geographical areas have the largest potential to grow in the future. I do believe that, there, we should not forget the newly industrialized countries. I believe that markets such as Spain and Greece in Europe; Mexico, Brazil and Venezuela in Latin America; Korea, Taiwan, Hong Kong, Singapore, Malaysia in Asia should be actively studied as to what strategies can be developed to open up these markets. And there I am very grateful to Coyne again for having given four key words. He said: "in order to open up a market you must first have legality, then respectability for gold, followed by the creation of infrastructure and then you can make your marketing efforts".

My last point is that one important aspect has not been discussed during the meeting: namely the impact of gold trade on other industries. I just give here the key words: the impact on transportation, on insurance business, on storage, on communications and last, but certainly not least, banking. I say this because I believe that the strength of the Zurich market is intimately connected with the capability of the Swiss banks to offer also related services.

I believe it would be interesting to discuss this aspect at a future

---

[5] See Jürg Niehans, *Gold Operations as an Instrument of Monetary Policy*, Part III, Chapter 24.

[6] See Luigi Stella and Fabio Torboli, *Aspects of Industrial Demand as to Jewellery: Case Study of a Typical Country*, Part II, Chapter 13.

Conference.

And just to conclude, three quotations by three French writers and philosphers on gold. The oldest is from Montesquieu, philosopher and economist who said: "L'or et l'argent sont d'eux mêmes inutiles, ce ne sont des richesses qui parce qu'on les a choisies pourront en être les signes".

But Molière, the poet, was more poetic and he wrote: "L'or donne aux plus laids un certain charme pour plaire et sans lui le reste est une triste affaire".

And the last quotation is from Jean-Pierre Renard who said: "Ce métal est un amour, un grand séducteur".

Part II

MEDIUM AND LONG TERM STRUCTURAL ASPECTS

Session chaired by Pierre Languetin

CHAPTER 9

THE MEDIUM AND LONG TERM STRUCTURAL ASPECTS FOR GOLD

LOUISE DU BOULAY and PETER D. FELLS

The purpose of this paper is to identify the long term trends which will influence the gold market. The analysis will be structured on production-supply and on demand according to a usual line of analysis.

1. Production and Supply of the Market Economies

Despite the sharply rising price, free world production has fallen by 26 percent since 1970 (Table 1).

South Africa has long been the dominant producer of gold and although her share of production is declining, still accounts for 70 percent of the total. The fall has occurred as part of a deliberate policy to take advantage of higher prices and mine the previously uneconomic lower grade ores. As a result, the average grade milled dropped from 13.3 gms per ton in 1970 to 7.3 gms per ton in 1980, lengthening mine life in most cases.

Tom Main's contribution will discuss future trends in South Africa in more detail, but we would like to make a few comments. The record price of 1979 and 1980 encouraged a surge of exploration activity, together with the re-examination of old mines, dumps and tailings in South Africa, as well as other countries. In addition to the announcement of some new projects, numerous expansion plans are being implemented, but several closures will also take place before the middle of the decade. The net result will be a levelling off in production at current levels until 1985-86 and thereafter, quite a sharp fall to about 350 tons by the end of the century.

This reduction will be brought about by a further fall in ore grades, as the mines plunge deeper, and the gradual exhaustion of certain mines. However, according to the comments of the ex-President of the South African Chamber of Mines at our gold conference in May, production will be less affected by variations in the price than in the 1970's. His projections based

Table 1 - *Free world gold production* (metric tons)

	1970	*1971*	*1972*	*1973*	*1974*	*1975*
South Africa	1,000.4	976.3	909.6	855.2	758.6	713.4
Canada	74.9	68.7	64.7	60.0	52.2	51.4
U S A	54.2	46.4	45.1	36.2	35.1	32.4
Brazil	9.0	9.0	9.5	11.0	13.8	12.5
Other Africa	44.4	44.6	42.3	44.8	43.6	40.0
Other Latin America	26.4	25.1	25.3	24.2	23.1	29.3
Asia	32.9	33.7	32.7	30.3	27.7	26.5
Europe	7.4	7.6	13.2	14.3	11.6	11.4
Oceania	23.8	24.7	39.4	40.7	39.9	37.4
Free world total	1,273.4	1,236.1	1,181.8	1,116.7	1,005.6	953.9

	1976	*1977*	*1978*	*1979*	*1980*	*E 1985*
South Africa	713.4	699.9	706.4	703.3	677.0	675.0
Canada	52.4	54.0	54.0	51.1	50.0	84.3
U S A	32.2	32.0	30.2	30.2	27.6	57.5
Brazil	13.6	15.9	22.0	25.0	35.0	75.0
Other Africa	39.2	41.4	34.2	28.3	32.2	43.2
Other Latin America	41.4	40.0	40.7	41.1	51.9	65.1
Asia	27.1	30.1	30.7	29.0	31.4	39.5
Europe	11.4	13.2	12.5	10.0	9.2	11.7
Oceania	38.9	44.0	45.5	39.8	32.8	63.5
Free world total	969.6	970.5	976.2	957.8	946.7	1,114.8

E - estimated.

on prices rising from a range of $395-495 in 1981 to $625-825 in 1990 in real terms showed much the same production profile.

Taken together, production in Canada, the US and Australia accounts for some 12 percent of free world output. In each case output declined steadily since 1970, as it became economic to mine lower ore-grades. The downward trend has now been reversed. Much publicity surrounded the new developments and re-openings inspired in these countries by the high prices of 1979 and 1980. Despite the recent price fall, we think that enough momentum exists to get most of the projects planned to 1985 underway, (although Australia, in particular, will be subject to cost pressures). This is in part due to the fact that a number of the proposed mines contain gold in association with base metals, which will tend to come on stream regardless of the gold price, but also the result of the fact that many of the projects are small, requiring little capital expenditure or development time.

In Canada, we have identified over 150 potential projects to 1985. The largest of these will produce 6 tons per annum, but most of the smaller ones will produce less than one-half a ton and have anticipated lives of less than 5 years.

In the US, production has almost halved since 1970, but has the potential to double as a result of the discovery of a large number of small prospects. Several of North America's major mining companies are responsible for the discovery and development of the larger projects identified. Most will be open-pit mines, and as in Canada, capital costs are low and lead times short. Other low-grade ore-bodies are known to exist, but of a complex metallurgical composition, and their development will proceed with caution.

Australian production peaked at 70 tons at the turn of the century. Four mines now account for 80 percent of production and a significant and rising proportion of the gold is produced as a by-product of base metal operations. We have identified 36 potential projects to 1985, ranging in size from 350 kilograms to 2.8 tons per annum. A number of large base metal deposits with significant gold values could start up after 1985.

The area which has responded most positively to the higher prices of the last decade is Latin America. This is because for the most part production is still obtained from small scale exploitation of alluvial deposits. The increase in Brazilian production from 9 to 35 tons has been particularly dramatic. The discovery of the Serra Pelada deposit in 1980 and the speed with which an army of *garimpeiros* stripped it of some 10 tons of gold, without infrastructure, capital, or modern equipment, demonstrates the potential response that can be expected to higher prices in countries with similar alluvial deposits. There are also indications that production may be expected to decrease significantly when prices fall. Brazil currently has only one hard-rock underground mine, open since 1835 and producing some 4 tons of gold. Government geologists have identified gold bearing reefs in Brazil, but the bulk of these will require proper long term mining development.

As the country lacks the domestic resources to develop the reefs, we see only a somewhat price elastic increase from alluvial production in the foreseeable future.

In the immediate future, two other countries are expected to increase their production sharply — Papua New Guinea and the Philippines. In both cases, gold is generally found in combination with copper and new projects are in an advanced state of development. The Ghanian government has ambitious plans for the further development of its country's undoubtedly promising gold mining industry. Like Brazil, lack of foreign exchange, and the problems faced by existing mines in obtaining spares and skilled operators make prospects of increase remote.

In summary, we think that whilst South African production will remain more or less flat until 1985, output in other parts of the free world will increase by approximately 170 tons. This will reduce South Africa's share of production to 60 percent of the total. After this, depending upon prices and the rate of development in other parts of the world, South Africa's share will fall further and faster, resulting in a very different pattern from that which we now have. Although the average size of the new projects being developed currently is small in comparison to the South African giants, the sheer numbers involved add up to a substantial quantity of gold. However, over the longer term, these mines will be less able to compensate for reduction in South Africa, resulting in a declining total output of gold to the end of the century.

The second point that we wish to make concerns working costs. According to work which we have done, based on 1980 data, the average cost of producing one ounce of gold in the free world was $190 an ounce — the average for South Africa being $176, and elsewhere $224. Although many of the new projects identified to 1985 are small and can be opened quickly, we reckon that most need prices of $300-350 an ounce (in 1980 money) to break even. As a result, the overall average cost of production worldwide will rise, placing a rising floor under future prices.

2. Production and Supply of the Communist Bloc

Both Russia and China have long histories of involvement in the international gold market. In Table 2 you can see our estimates of annual net trade with the Communist Bloc covering the activities of not only Russia and China but also other East European countries and North Korea, where appropriate. It can be clearly seen that there has been a considerable variation in the levels of sales and purchases over the period shown.

Before the Communist revolution China traded heavily and during the Civil War leading up to the Communist take-over, annual imports of over 90 tons were recorded by the BIS, apparently for private hoarding. In 1952, the

Table 2 - *The supply of gold* (metric tons)

	Non-communist world mine production	*Net trade with Communist sector*	*Net official sales (+)*	*Net official purchases (—)*	*Total*
1948	702	—	—	369	333
1949	733	—	—	396	337
1950	755	—	—	288	467
1951	733	—	—	235	498
1952	755	—	—	205	550
1953	755	67	—	404	418
1954	795	67	—	595	267
1955	835	67	—	591	311
1956	871	133	—	435	569
1957	906	231	—	614	523
1958	933	196	—	605	524
1959	1,000	266	—	671	595
1960	1,049	177	—	262	964
1961	1,080	266	—	538	808
1962	1,155	178	—	329	1,044
1963	1,204	489	—	729	964
1964	1,249	400	—	631	1,018
1965	1,280	355	—	196	1,439
1966	1,285	—67	40	—	1,258
1967	1,250	—5	1,404	—	2,649
1968	1,245	—29	620	—	1,836
1969	1,252	—15	—	90	1,147
1970	1,273	—3	—	236	1,034
1971	1,236	54	96	—	1,386
1972	1,182	213	—	151	1,244
1973	1,117	275	6	—	1,398
1974	1,006	220	20	—	1,246
1975	954	149	9	—	1,112
1976	970	412	58	—	1,440
1977	972	401	269	—	1,642
1978	979	410	362	—	1,751
1979	961	199	544	—	1,704
1980	943	90	—	230	803

Definition of official sales has been extended from 1974 to include activities of government controlled investment and monetary agencies in addition to central bank operations. This category also includes IMF disposals.

Chinese market was closed and since then the Bank of China has been the sole Chinese presence in the market. China has recently become a member of the IMF and has reported official gold holdings of 398 tons. Given her past history, it is likely that other stocks of gold are held in the country, both in official and possibly private hands. In addition, we estimate that China currently produces somewhere between 25 and 45 tons of gold per annum and has the potential to double this.

However, most of the existing mines are small and old, as is their equipment. Although there are reports that gold has been discovered in many provinces and increased production is a government priority, we think that at present, China lacks both the financial and technical resources for large scale mine development. We believe that the central bank takes most of the gold produced into reserves. For the future, China could therefore increase her production and will remain a potential source of gold to the market.

Russia also remains somewhat of an enigma due to the secrecy that has shrouded all her activities since the Revolution. However, apart from periods of total withdrawal, she has been a source of substantial quantities of gold to the market. Russia's current gold mining industry took shape under Stalin and output doubled during the 1930's. It has doubled again since, due, amongst other developments, to the opening of the giant Muruntau mine in 1969. Total production now stands at an estimated 300 tons per annum. As in China, the expansion of Russian gold output is a government priority, but owing to severe technical and climatic problems, together with the widespread inefficiency and failure to meet targets which plagues Russian industry generally, progress is slow.

It is assumed that Russia sells gold to the West in direct response to her needs for foreign exchange. The quantities sold have varied from a peak of 489 tons in 1966 to nil in other years. We have made some attempts to rationalize the pattern of Russian sales in order to make some predictions for the future. Russia obviously sells gold because she produces it and it is one of the few exportable goods that can be sold quickly, relatively anonymously and without the political and security repercussions attached to sales of arms and energy materials. Gold appears to be regarded as an asset of last resort, when all other things have failed.

Two readily observable factors seem to show some correlation with Russian gold sales from year to year, but there are nevertheless inconsistencies.

1) her hard currency balance of payments deficit (current account)
2) her level of grain imports.

In 1965, Russian grain imports quadrupled, coinciding with her highest ever gold exports. During the period 1966-1970, grain imports were low and no gold was sold. In 1972, grain imports reached previously unprecedented levels and gold sales picked up. From then until 1975, there is a relationship between the trends in the trade balance, grain-purchases and gold sales. The movements of 1976 and 1977 are less easy to explain. However, in both 1979 and 1980, grain purchases rose, the trade deficit fell and gold exports halved from year to year. One explanation for this is of course the sharply higher gold price, but it is possible that oil revenues or some other unknown factor improved the trade balance. During the first half of 1981, the Russians' hard currency deficit was already in excess of the totals for 1977 and 1978 and the prospects for the harvest were poor. A Kremlin memorandum warned of food rationing in the summer. From these facts and the lower grain prices, we anticipated higher gold sales and this has proved to be the case.

Although in the case of Russia, nothing can be predicted with much confidence, we think that there are reasons for supposing that she will continue to sell gold in the foreseeable future and probably at a higher rate than in 1979 and 1980. The reasons are as follows:

1) From domestic production of 300 tons per annum at least 50 tons and maybe more are absorbed each year into both industry and stocks, leaving 200-250 tons available annually for export. Additional stocks are, however, available in case of emergency.

2) After World War II the Soviet economy grew faster than that of the US, but this situation reversed itself in the 1970's. There are growing signs that the Soviet system is not suited to the operation of a complex modern economy and denunciations of mismanagement and production short-falls have become more frequent. Apart from the arms industry, no area of Soviet production is booming, leaving gold as one of the few readily available commodities for which the West will pay.

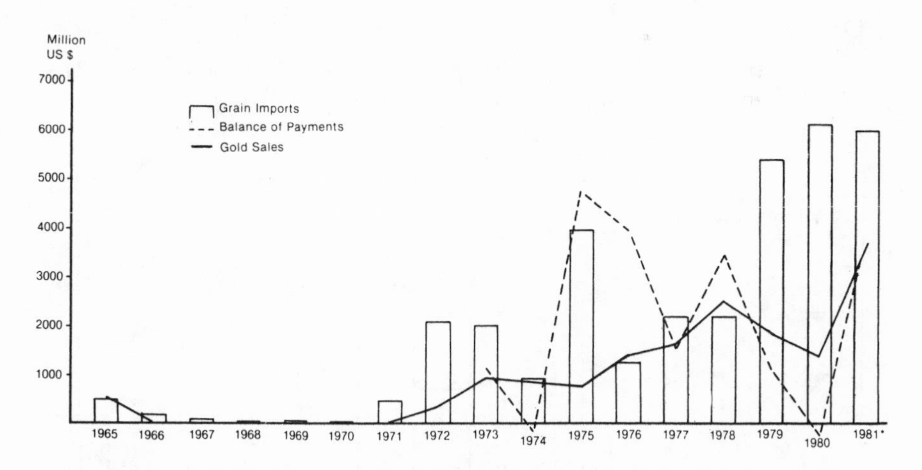

Fig. 1 [a] USSR grain imports, gold sales and current account hard currency balance of payments deficits.

* Balance of Payments: 1st half only. Grain imports and gold sales full year estimate.

3) It is currently estimated that oil exports account for well over 50 percent of Soviet hard-currency earnings. US intelligence sources estimate that in the future, oil exports to the West will decrease as the rate of production drops and domestic consumption rises. In addition to this, oil prices have stabilized and Western demand has fallen. As a result, revenue from Russian oil

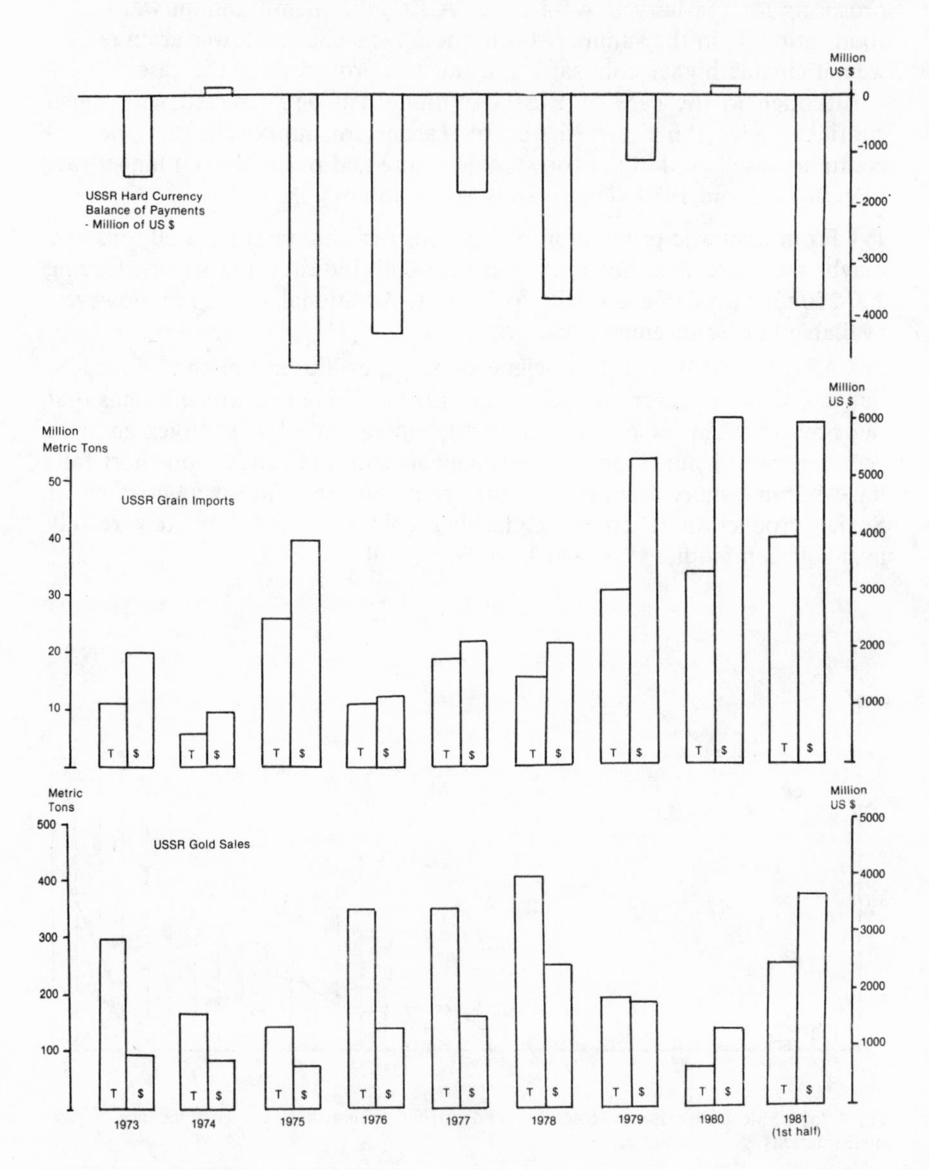

Fig. 1 [b] USSR hard currency balance of payments; USSR grain imports; USSR gold sales.

exports was said to have dropped 20 percent in the first half of 1981. Despite the development of other energy resources, the export of such things as gas will probably not help offset declining oil revenues before the middle of the decade, at the earliest.

We therefore assume that as Russia's needs for foreign exchange seem to be directly related not only to levels of grain imports, but also to revenue from oil sales, then as oil exports tail off, dependence on gold sales to the West will increase. The quantities sold will vary in accordance with internal circumstances and the gold price in any one year, but seems likely to be in the order of 200-250 tons per annum.

3. The Official Sector

1980 looks like a watershed in the monetary history of gold; for the first time for decades no monetary authority of importance, national or international, committed itself in word or deed to influencing the world gold price.

Until 1968, the world's major monetary authorities combined to maintain the gold price at $35 per ounce. This required substantial purchases of metal from the private sector up to the mid-sixties, after which growing jewellery and industrial consumption of gold, accelerating inflation, the increasing international overhang of dollars, and speculation on an ultimate breakdown of the system, all pushed up private demand dramatically. In March 1968, the central banks gave up the struggle to maintain the $35 price and ceased selling.

As it happened, it was then necessary to resume purchasing in the early 1970's to prevent the new, free market, price falling below $35 under the weight of speculative liquidation. Thereafter the US and the IMF fought an increasingly lonely battle, by means of propaganda, intimidation of other monetary authorities and gold sales programmes to contain the rise in the price of gold and prevent the resumption of its use as an international monetary medium.

Happily, that is now all in the past, and gold is free to be traded and used by the central banking community as it sees fit. We will leave others to debate whether this is theoretically desirable or not, but whatever the academic verdict, the present laissez-faire regime is likely to persist, for much the same reason as floating foreign exchange rates are likely to persist. That reason is political reality; no single authority today, national or international, is sufficiently dominant to regulate a system. The world of the 1980's is very different from the Bretton Woods' world of 1946, when the economies of Europe and Japan lay in ruins and primary producing countries were still colonies.

The freedom of 1980 was used by the central banking community to add

230 tons of gold to its collective reserves. We would expect the official sector to remain judicious net buyers, on average, in the years ahead.
The reasons are:

(i) Those countries which have experienced a rapid build-up in reserve holdings in recent years — principally some of the OPEC countries and some newly industrialized countries — have a disproportionately small gold component in their reserves, and will wish to redress the imbalance.

(ii) Because it includes several countries of comparable economic importance, the EEC regional economic grouping is unlikely to adopt any one member's national currency as an intra-group reserve. Gold is therefore likely to remain an integral part of the EEC payments accounting system. The use of gold by such an influential group will encourage interest elsewhere.

(iii) While the US dollar will remain the principal reserve asset in the Western hemisphere, the US itself will avoid selling down its own gold stock if at all possible, because the acceptability of its own currency outside the Western hemisphere will remain suspect. Its gold stock will be viewed as essential backing for dollars held overseas.

(iv) Gold producing centres, even within the Western hemisphere, are likely to continue adding some of their output to their reserves, where they can afford to do so.

We would expect, therefore, the central banking sector will be on balance a net purchaser of gold in the years ahead. That is not to suggest it will be a rash or headstrong buyer. On the contrary, given the natural risk aversion of central banking officials, we would expect the sector to be a stabilizing influence on the price, purchasing in a weak market and abstaining, or perhaps even selling, if the market begins to soar. The embarrasment of US official sales of gold at an average $255 per ounce in 1977-79, and Indonesian official purchases in 1980 at a time when the gold price was over $600 per ounce, is presumably still fresh in the mind.

4. Consumption Demand

Having covered the supply side, let us turn to demand, which is summed up in Table 3.

Gold is used for a great many decorative and industrial purposes and in Table 3 we have a breakdown into the major end-use categories. Our figures are put together on the basis of the country in which the gold was fabricated rather than the one in which it may ultimately be sold. It may be seen that whilst the bulk of fabrication takes place in the industrialized world in years of good demand, the markets for jewellery and coin are much more evenly

Table 3 - *Gold fabrication* (metric tons)

	1970	*1971*	*1972*	*1973*	*1974*	*1975*	*1976*	*1977*	*1978*	*1979*	*1980*
Fabricated gold in developed countries											
Carat jewellery	500	553	702	428	278	317	471	540	592	551	270
Electronics	89	86	105	126	91	66	75	76	84	92	79
Dentistry	55	59	61	64	54	58	73	78	85	82	60
Other industrial/decorative uses	58	62	65	67	64	55	59	59	70	69	61
Medals, medallions and fake coins	29	33	32	19	11	10	20	22	21	16	18
Official coins	32	44	44	36	209	221	145	125	256	242	164
Total	763	837	1,009	740	707	727	843	900	1,108	1,052	652
Fabricated gold in developing countries											
Carat jewellery	566	511	297	90	—53	206	464	463	415	186	—150
Electronics	—	—	—	1	1	1	1	1	2	2	2
Dentistry	4	4	5	4	3	4	4	4	4	4	2
Other industrial/decorative uses	4	6	5	4	3	4	4	5	7	6	3
Medals, medallions and fake coins	25	19	9	2	—4	11	30	29	29	17	—3
Official coins	14	10	19	18	78	30	37	17	31	48	15
Total	613	550	335	119	28	256	540	519	488	263	—131
Total fabricated gold	1,376	1,387	1,344	859	735	983	1,383	1,419	1,596	1,315	521

split between the developed and developing world. The negative figures indicate that more gold was dishoarded on to the international markets than was bought, whilst any manufacturing was from local recycled gold. This method of reporting demonstrates that when prices are high, owners of jewellery in developing countries are apt to cash it in in substantial quantities.

Trends in jewellery sales are hard to tie to any single factor. In the first place, motives for buying jewellery vary widely — for instance, in Europe, gold jewellery has historically been worn as a decoration and status symbol whereas in the Middle and Far East, it is impossible to distinguish between the buying of gold for decorative or investment purposes. Secondly, a great deal of jewellery is fabricated for export — as in Italy — making the comparison of fabrication and such things as GNP growth in any specific country thoroughly unsatisfactory. Thirdly, whilst jewellery in the developing countries is generally simple and sold at a low mark-up to the gold weight, in the industrialized world, a mark-up of up to 600 percent over the gold content is not uncommon.

However, even the most cursory glance shows that jewellery fabrication and the gold price has a crude inverse relationship. For example, when gold was still selling at the official price in 1970 and 1971, jewellery fabrication exceeded 1,000 tons per annum. Sales then declined to 1974 as prices rose, but by 1975 buyers had apparently adjusted to the new levels. A similar trend can be observed more recently and we believe that this pattern will continue for the future.

For several reasons, we think it likely that potential jewellery usage, in the sense of the net amount of gold demanded at any given price for the metal, will be on a rising trend:

(i) Despite the current economic gloom, real personal incomes are more likely to resume rising in the longer term. Indeed, for many of those in employment, they have not stopped rising. In many industrialized countries, the markets for such things as cars and other consumer durables are almost at saturation point. Under these circumstances, and particularly in countries like Japan where space is at a premium, small luxury items should gain in popularity, particularly as, in the case of gold, they are perceived to have an intrinsic value of their own.

(ii) Through publicity and a growing number of retail outlets, a greater number of people are aware of gold and its uses.

(iii) There are signs that attitudes to jewellery in the West may become more like those of the Middle and Far East. Consumers in some industrialized countries, possibly as a result of the publicity and attempts to sell jewellery as scrap during 1980, are beginning to look for jewellery at prices which are noticeably more related to the gold content. Semi-fabricators and manufacturers, who are obviously keen to boost sales, are expressing interest in opening their own retail outlets or in selling direct to big chain stores and mail order firms. The result should be that the consumer's dollar buys more gold and less fabricator and distributor mark-up.

This brings me on to the subject of coins and small bars. Sales of bullion coins have increased remarkably since 1973. In part, this growth reflects the greater availability of coins during the past decade — starting with the Kruggerrand, then the Maple Leaf and even the Russian Chervonetz. The success of the Kruggerrand is undoubtedly due to the marketing efforts of the Chamber of Mines, plus their ready adaptability to the perceived needs of the market in producing smaller denomination coins. In addition to coins, for the first time last year we became aware of a growing demand for small bars, ranging in weight from 1 to 100 grams. These are being made mainly in Europe, North America and Japan, but are on sale in most parts of the world. The tonnage involved could amount to half that for official coins. The choice between coins and bars usually depends upon the purchase tax charged in each particular country.

A mixture of motives has promoted coin and small bar purchases. They include a better-value-for-money method of purchasing ornamentation — this would apply, especially to the very small denomination coins; a convenient form of tax evasion; and small-time investment or speculation. In the former instances, the gold is unlikely to be sold in quantity in response to a price rise, and even in the case of investment, the usual practice of physical possession discourages trading.

The complexity of motives underlying the purchasing of coins and small bars, and the comparatively recent emergence of the phenomenon, makes it very difficult to postulate future trends with any confidence. We take comfort from the fact that in the extremely adverse economic climate of 1981, and with a gold price still averaging $459 per ounce, coin purchases stayed above 200 tons and suggest this could be a 'fail-safe' level to assume for the future.

There are signs of a continued fall in the use of gold for electronic purposes. In the electronics sector plating weights have been further reduced and some switch to other alloys has taken place. As a result of the high prices, most companies have redoubled their research effort into such economies. Although the number of electronic components being manufactured is expected to grow, technology is changing so fast that it is hard to tell whether this will compensate for continued reductions in gold use per unit.

For dental alloys, as with jewellery, the high mark-up charged by wholesalers and dentists alike has proved detrimental. There are reports that the new, lower gold content alloys do not bond well with ceramics and that the costs of fabricating other non-gold alloys result in the use of gold working out cheaper. Nevertheless, we can detect no change in the declining statistical trend.

One final comment concerns the recovery of scrap. Encouraged by the rise in prices, many industries improved their recovery and achieved lower levels of process scrap generation by more efficient gold usage. The high prices also boosted the collection and recovery of low-grade old scrap. Both trends will tend to contribute towards a reduction in the demand for new gold for industrial purposes in the future.

5. Investment Demand

One of the functions of money is as a temporary abode of purchasing power which can be immediately mobilized. One of the functions of gold is to serve that function at times when there is apprehension that paper currency will fail to fulfill it.

Apprehensions can be very fickle, and they have been largely responsible over the last fourteen years for marching the gold price to the top of the hill and marching it down again. Little purpose will be served by conjecturing what they will do this year or next. It is possible, nevertheless, to identify some longer term influences:

(i) The demand for gold in geographical areas of political turbulence is well attested, both by experience and common sense. There is no point in amassing local currency if the issuing authority is likely to disappear, or if one may wish to take refuge in a country where it is worthless. Our guess is that, on balance, over the years, there will continue to be a positive net demand for this purpose.

(ii) Swiss bankers have long recommended that clients hold a minimun fraction of their portfolio in gold, virtually regardless of the prevailing economic climate. Of late, portfolio management theory has tended to give academic respectability to the policy, on the grounds that it reduces variability of return.

(iii) Gold holding in a number of countries, including the US, UK and Japan, has become legal only comparatively recently, and the market in these cases is almost certainly far from saturated.

(iv) Much more controversial is the scope for a long term demand for gold as an inflation hedge. Logically this question should be split into two:

a) Will inflation continue to be a problem?

b) If it does, will gold be bought as an inflation shelter?

Topic a) merits a paper in itself. Personally, I feel that inflation is endemic to social democracy. To achieve and retain power in a state dedicated to social democracy (and virtually all the Western democracies are now so dedicated) a politician cannot afford to offend the various vested interests in social security by cutting their benefits. Just to illustrate this statement, though one could choose many other examples and I do not wish to be invidious, it is clear that the state pensioner now enjoys tremendous political clout in many countries, and no government can afford, in electoral terms, to let the real value of pensions slip. Yet this imposes a burden on the working population which, for demographic reasons, is bound to increase. At the same time a government courts across-the-board unpopularity if it attempts to cover the costs by heavier taxation. Budget deficits appear inevitable.

Monetarism teaches us that budget deficits need not be inflationary if they are financed by long term borrowing. But this can only be a short term palliative. Crowding out private sector borrowing must depress trade and industry and create unemployment. In itself, this may be enough to persuade the government of the day that it must reinflate prior to the next election. But even if deflation is doggedly pursued, the decay of the private sector must erode the tax base and hence exacerbate the budget deficit, in a vicious cycle. Sooner or later, the siren voices prevail and recourse is made to the printing press.

b) Is gold an inflation hedge? Professor Jastram's fascinating book *The Golden Constant*, indicates that historically gold has appreciated in a deflation and depreciated in an inflation, measured by real purchasing power. I hope he will forgive my recalling the famous dictum about lies, damned lies and statistics. What I think the record shows is that when a monetary authority is determined and able to maintain the gold parity of its currency, the conclusion inevitably follows. The most recent and familiar example is the period 1934-1968. In periods when the onus of convertibility becomes intolerable, and the gold price is no longer manipulated, the metal indeed becomes an inflation hedge.

6. Conclusions

We have suggested:

1) South African gold production will remain on a plateau for the next five years, but halve by the end of the century — regardless of price.
2) The survey of new projects and expansions which we undertook last year indicated that output in other parts of the free world could increase by about 170 tons by 1985. We believe that despite the fall in prices since then, most of the mine projects planned will come on stream, but that alluvial production in the developing countries such as Brazil will prove more price-sensitive.

3) The supply of gold from the Communist world will rise gently, reflecting continuing maximum Soviet exports.
4) Central banks as a totality will be inclined to add to their gold stocks, but purchase unaggressively.
5) Jewellery demand, at any given real price of gold, will continue to rise.
6) Purchasing of coins and small bars will, at a minimum, be maintained at present levels.
7) There will be a moderate decline in industrial usage of gold for some time.
8) Political and economic stresses will on balance encourage private investors to add to their gold holdings.

It would be quite misleading to translate these thoughts into figures, but you will not need a very sharp pencil to work out why we at Consolidated Gold Fields are continuing to fund a large exploration programme for this most alluring of metals.

CHAPTER 10

FACTORS AFFECTING PRODUCTION AND COSTS IN THE MEDIUM TO LONG TERM IN THE GOLD MINING INDUSTRY

THOMAS R. N. MAIN

1. Price Fluctuations

My brief is to consider the factors affecting production and costs in the medium to long term in the gold mining industry. In doing so I intend to concentrate on the gold mining industry in South Africa which is far and away the major gold producer at the present time. However, much of what I have to say on the South African industry has applicability to the gold mines in other parts of the world.

A distinguished historian once said: "From 1886 the story of South Africa is the story of gold". That was in 1941. Forty years later this statement is as true as it was then. The story of South Africa is still the story of gold. Today, when the gold price is high the economy booms, when it is depressed the economy slows down and pessimism amongst economic forecasters becomes the new conventional wisdom.

It therefore comes as a surprise that the demise of gold has so often been predicted since the establishment of the industry. An example of such faulty predictions was in 1930 when a government mining engineer predicted that the peak of gold production would be reached in 1932. The subsequent increase in the dollar price of gold prolonged the life of the industry well beyond its confidently predicted expiry, but continued prophecies of decline over the years show that the gold mining industry is not immune to crises. It has survived, and prospered remarkably, but as the future is inherently uncertain, particularly, it seems, in the world of the late 20th century, no-one can foretell with absolute confidence the future of the noble metal except to acknowledge that its attributes have been accepted for so many centuries that it is highly unlikely that the position will alter in the foreseeable future.

Paradoxically, the most recent era of prosperity for the gold mines has been a reflection of global crisis. The establishment of the two tier marketing system in March 1968, was both a symptom of the instability of the international monetary system, and the beginning of the gold mining boom. International political uncertainty, energy crises, and global inflation have all made their contribution towards increasing the demand for gold and thus raising its price. And, it is of course the gold price which is the most important factor in any consideration of future production.

There have been two outstanding features of the history of gold since the beginning of the two tier system and the final collapse of Bretton Woods in 1971. These are the sharp rise in its price and its substantial volatility. The average price rose from less than $36 per fine ounce in 1970 to $613 per ounce in 1980. Since then there has been a sharp downturn in the price, although at an average of $460 an ounce in 1981 it was still high by the standards of only some years ago. It is this volatility of the gold price which makes it so difficult for the mining industry to plan for the future. Before mid-1978, the price usually fluctuated on a daily basis within a range of $1 to $2 per ounce. Since then it has generally moved by many times that amount as witness its decline by $113 an ounce over one 24-hour period in early January, 1980.

Both these phenomena, the high price of gold as compared with the levels of the early 1970's, and its volatility, can be explained in similar terms. Perhaps the single most important factor was the sharp increase in the speculative demand for gold. This in its turn was a response to the deep uncertainty about the future which political and economic events seemed to dictate. Accelerating world inflation reduced the return on traditional investments such as equities, and enhanced the attractions of gold as a store of value. The threat to world stability posed by political disorder in the Middle East further contributed to the world demand for gold. These factors also accentuated gold's volatility. Speculative demand is inherently more volatile than the more traditional forms such as jewellery and industrial demand for gold. It is potentially, although not necessarily, destabilizing. This seems to have been so with gold. Rising prices led to expectations of still higher prices. Such expectations may, before 1981, have been in the right direction, but they also tended to be excessive and resulted in exaggerated price movements.

Price volatility has also been enhanced by the growth of futures markets in gold, which speculators have found a more attractive area for high returns than spot markets. General economic instability throughout the world, reflected in frequent changes in interest rates and currency values, has also led to greater variability in the gold price. In addition, the large number of new participants with no previous experience in the gold market, expectations about the monetary role of gold, and sharp changes in the amount of gold sold by the Soviet Union, have all added to the volatility of the gold price.

2. Price of Gold, Uncertainty and the Mining Industry

It is clear that the gold mining industry must learn above all how to cope with uncertainty. Most obviously, the price of gold determines the grade of ore mined and therefore the output of gold as well. South African gold mines are legally required to mine to the average value of each mine's published ore reserves. Such reserves are determined by the pay limit, i.e. the minimum quantity of gold in a ton of rock which will produce enough revenue to cover the costs of mining, processing and marketing the product. When the price of gold rises the pay limit declines, lower grade ore is mined, and previously marginal mines acquire a new lease of life. This also affects the output of gold, which declines in volume as the grade which is mined falls. At the same time the profitable life of a mine is extended.

The obvious problem the gold mines have to face is how to plan their production to offset the disadvantages of the volatility in the price. Although it can realistically be expected that the gold price will rise in the long term at a rate which is not less than the rate of global inflation, fairly small fluctuations in price can have enormously costly effects. A price rise which is not sustained can lead to expansion programmes involving huge losses that can never be recovered.

The problem has partly been met by some mines reviewing pay limits far more frequently than before, which means that they can now mine to more or less the average value of the ore reserve from quarter to quarter, or from month to month, as the case may be. More fundamentally, the production planning of the gold mining industry is based on a blend of conservatism and a definite optimism about the long term prospects for gold. The industry is convinced that the price of gold will rise in real terms over the next decades, whatever the temporary downturns may be. It is this conviction that accounts for the immense expansion programmes that have been undertaken during recent years. Existing mines have been expanded, new ones have been opened, and the lives of marginal mines have been extended through the mining of lower grades of ore. To a very large extent this involves an act of faith. If the gold price is depressed in the long term then there is no way in which the industry will avoid losses. But this is what private enterprise is all about: the willingness to place one's money on what seems the most realistic assessment of the future and to accept whatever gains and losses that may bring.

It is this growing realization of the problems experienced by the volatility of the price which has led many producers to seek to protect their earnings by hedging their production on the futures or forward markets. This is an aspect receiving the careful attention of certain South African producers and is likely to be an important aspect of mining policy in the future.

3. Reasons for Long Term Optimism on Gold

While uncertainty is an inherent element of all production planning, which can at best be minimized, but never eliminated, the mining industry believes that there are excellent reasons for long term optimism about gold. This is closely linked with a basically favourable appraisal of the development of the world economy. The most important components of the demand for gold, viz. jewellery and industrial demand, have been depressed during the last few years partly because of the low level of world economic activity. Should the global economy recover and with it real disposable incomes, particularly in the developed countries, then it can be expected that these areas of demand for gold will return to previous levels. In the case of jewellery demand relatively high and variable prices have led to a decline but the current lower prices and rising disposable incomes, at least in some parts of the world, such as the Middle East and the Far East, can be expected to result in an improvement in the jewellery demand for gold as was seen in 1981 compared with the previous two years.

There is also another component of demand which should contribute to a higher gold price although its course is much less predictable than those just discussed. This is the monetary demand for gold, and its growth is obviously linked in the closest way with the future of the international monetary system. There has been much talk in the United States about a return to the gold standard, mainly by "supply-side" economists like Arthur Laffer who believe that a return to gold is the only guarantee against the debasement of the dollar, both at home and abroad. Even if there is no return to the gold standard in its classical form there is little doubt that the efforts to demonetize gold as an international reserve asset have failed. Gold reserves valued at market prices remain the largest proportion of world reserve assets held by central banks. The European Monetary System has assigned a vital role to gold by using it as part of the basis of valutation for the ECU. It is likely that gold will continue to play an international monetary role of some importance, even if a return to a fullblown gold standard is unlikely.

The future of global inflation will naturally have crucial implications for gold. If it accelerates uncertainty about the future will increase the attractiveness of gold as an asset less susceptible to erosion of its real value. But speculative demand will be given a new lease of life and price volatility will be accentuated. If, however, the oft-repeated enunciations of the governments of the developed world that inflation is the main enemy are to be taken seriously, then it is likely that the rate of increase of global inflation will gradually decelerate. The price of gold will rise less rapidly, but its variability will also be reduced because of the decline in speculative demand. In such circumstances it is probable that the price of gold will rise fairly continuously, even if less spectacularly than during recent years, sustained by the steady growth in the more stable components of demand for gold, namely, jewellery, industrial and the monetary demand for gold.

It cannot, however, be expected that volatility in the gold price will be eliminated. Even if world inflation were to be subdued and brought under control, it would be naive to expect that the years ahead will be devoid of major international crises, such as the Soviet invasion of Afghanistan and the troubles in Iran. In such trying times the price of gold, as a classic hedge against uncertainty, is almost bound to rise. A slow but steady increase in the gold price in the long term would therefore be quite compatible with short term fluctuations which could inflict enormous losses on the gold mining industry if they were completely unanticipated.

4. Planning of Gold Production and Cost Behaviour

This is why long term optimism must go hand-in-hand with a basic conservatism about the planning of production. The gold mining industry has tried, as far as human foresight can allow, to safeguard itself against the vagaries of time and chance, however tempting it may be to respond with heady optimism to sharp and, apparently, continuous increases in the gold price. The assumption in the gold mining industry is that the price of the metal will be substantially less than the staggering heights it reached in 1979 and 1980. This has been a lesson learnt only by experience and at some cost. There has been capital expansion in the past, undertaken in response to an escalating gold price, which had to be cut back when gold came tumbling down.

The present policy of the industry is therefore tough-minded, rather than light-headed. It is calculated that when the current expansion programmes are completed they would at least break even at a price of $350-400. Totally unforeseen events could push the price even below this conservative limit, but risk-taking is inherent in private enterprise. What is clear is that the industry has not abandoned its sense of proportion simply because of events during 1979 and 1980. In the light of what happened to the gold price in 1981, it is a policy which has been amply justified by the test of experience.

During the next few years, South African gold production will probably remain much the same as it is now. Production by the end of the century will depend on so many imponderables that one is largely reduced to guesswork. Conservative projections suggest that total gold output will remain at about 700 tons per annum in the 1980's, varying perhaps by two to four percent per annum. Forecasts beyond the end of the century indicate a gradual drop to about 350 tons per year. Such projections in the past have proved to be wrong on so many occasions that any projection beyond five years becomes quite meaningless.

These projections are naturally based on certain assumptions about supply and demand. The latter has been discussed at some length. The question remains what will happen to supply, as it is manifested in the costs of gold mining. The overriding consideration here is that both operating and capital costs have risen very rapidly indeed over the last decade. While the South African consumer price index increased by 228 percent during this period, costs multiplied six times. Operating costs have been rising at a rate of 16-20 percent per annum, and capital costs have been going up even more dramatically.

Labour costs account for about 50 percent of the direct costs of gold mining. The most impressive increase of all has been the rise in the wages of the unskilled labour force. This has resulted over the last decade in a most conspicuous narrowing of the skilled/unskilled wage gap. The average wages of unskilled migrant workers rose by 280 percent in real terms despite a rate of inflation which was in double digit figures for most of the decade. In addition to direct wage payments substantial funds have been spent on improving and extending the accommodation and other facilities of the resident work force. Over the last three years such expenditure has exceeded $350 million.

Partly the rapid escalation in working costs has been a straightforward manifestation of a rising rate of inflation. But there have been years when working costs have significantly exceeded the rate of inflation, without a compensating rise in the gold price. Historically, the gold mining industry has been extremely conscious of costs. It clearly had to be during the era of the fixed gold price. But even when the price was free to respond to market forces its volatility and unpredictability have not permitted rising costs to be treated with levity. Indeed, the pressure of escalating costs is such that if there is no marked recovery of the gold price over the next few years, marginal mines will again require state aid as is already the case with a few mines. As for non-marginal mines, if the gold price does not recover, it seems inevitable that they will have to reconsider existing capital expansion programmes. So far the reaction to the lower price in 1981 has been some peripheral reductions in less essential capital expenditure but the major developments are still well on course. I firmly believe, however, that the present lower gold price is a depressive phase in the cyclical gold price and that the developments in the gold mining industry will prove to have been well timed for the next upward movement in the gold price.

What applies to operating costs is equally true of capital costs. They have been rising steeply and the total amounts involved are immense. As an illustration of this problem there is the Elandsrand mine which was started in 1975 and cost about $310 million to reach full production. The lower gold price has resulted in the generation of insufficient cash from current operations to fund the rapidly rising capital cost requiring the raising of additional funds by means of a rights issue.

A further example of the huge capital costs involved is the No. 1 shaft of the Western Deep Levels mine which is the largest expansion programme on a single mine in the industry. It has been estimated that the total capital spending involved in this development would be about $1 billion.

These are only two illustrations of what has been a conspicuous feature of South African gold mining in recent years, viz. the immense expansion of capital spending, both on new projects and on the improvement of existing capacity. This has been a direct consequence of the higher gold price, which has made such investments profitable even when production costs are rising at annual rates well above the rate of inflation. Although most of these expansion programmes are based on a conservative gold price, it is likely that if the price were to fall substantially further than existing levels some capital spending projects will have to be reconsidered. It is of course the gold price some years hence which is the vital consideration as it takes a major new capital project about five years to come to fruition, an aspect which is often overlooked when assessing future prospects for the industry.

A feature of the rise in capital costs has been the trend towards the merging of individual gold mines into larger units. This procedure enables the capital cost of new projects to be offset against the profits of existing mining operations for tax purposes and permits more efficient utilization of existing mining infrastructure.

In spite, therefore, of the relatively high average gold price in 1979 and 1980, the costs associated with capital spending are so high and the uncertainty of what the future will bring so unavoidable, that the recent commitment to expansion can only be attributed to a deep-rooted confidence in the long term future of gold. There is nothing inevitable about this at all. Huge profits in one year, which are the usual focus of popular attention, can be more than counteracted by the negative profits in another which a depressed gold price can bring. Gold mining must be seen as decision-making in the face of uncertainty. It is not simply a question of taking it easy and waiting for the profits to flow in. The risks involved are high and the outcome cannot be known. The words of Keynes are particularly apt here: "The hypothesis of a calculable future leads to a wrong interpretation of the principles of behaviour which the need for action compels us to adopt, and to an under-estimation of the concealed factors of utter doubt, precariousness, hope and fear". But these decisions are based on carefully assessed trends and we believe that the gold mining industry has a dynamic potential.

In an industry where rising costs are a constant threat to a viable future attempts have been and are being made to keep them in check by technological improvements and the more efficient use of the labour force. Mechanization could in principle improve efficiency and extensive research is being undertaken by the industry in this field. Natural constraints on gold mining in South Africa, such as the great depth at which mining has to be carried out and the narrowness of the orebody, have inhibited progress towards economically viable mechanization. Consequently the industry will continue to be labour intensive although there are indications of the industry's intensive research effort yielding important technical developments which will have a vital bearing on the industry's cost structure in the future.

The future of gold mining in South Africa is very much dependent on the use that is made of its available labour. There is at present a severe skills shortage in the mining industry. Clearly the solution to this problem is the advancement of unskilled workers into more skilled positions and this development is now very much a reality.

One development in the field of labour has at least reduced uncertainty about the future. The fears of an interrupted supply of unskilled workers have proved groundless for improved wages have increased the attractions of mining and an increasing proportion of the labour force has come from the South African geographical region. At the same time there has been increased stabilization of the labour force with migrants being offered strong incentives to make mining a career. The supply of unskilled labour is therefore no problem to the gold mining industry certainly at the present time.

5. Observations on Countries Other than South Africa

South Africa is much the largest gold producer accounting for some 70 percent of production of the market economies in 1981, and around 57 percent of supply including sales from the centrally planned economies. Of the other producers the largest is the Soviet Union with an estimated production of around half that of South Africa. Regrettably little information is specifically available about production or costs of gold mining in the Soviet Union. With insufficient definitive data and little knowledge about the efficiency of Soviet production techniques it is virtually impossible to even hazard a guess as to future production levels. Nevertheless the impression generally gained is that output is likely to show some modest increase in the next few decades. Actual sales of gold by the Soviet Union is another matter entirely and probably even more difficult to project than production.

The rise in the gold price in the last five years and even at existing lower levels has resulted in substantial activity wherever gold is known to be. In China, Canada, the United States, Brazil, Ghana, Zimbabwe, the Philippines and Australia — the bug has bitten. Indeed so dynamic has been progress that it has been exceptionally difficult to keep track of the spate of new mines and projected developments. In all these cases output is relatively small and certainly does not in any way begin to compete with the major producers. Nevertheless, in total, additional supplies from these producers will help to offset the fall in output from South Africa which has taken place over the last decade.

Inevitably the seemingly endemic problem of inflation with concomitant increases in operating and capital costs have had a deleterious impact on all gold producers — not only those in South Africa — negating many of the advantages of the improvement in the gold price since the beginning of the 1970's. The impact not surprisingly depends on the nature of the producer. Underground mining is usually considerably more expensive than open pit mining, which also increases in cost with depth. Furthermore, costs vary according to the methods of mining, concentration and smelting required. In the case of by-product gold recovery costs are usually even lower because most of the process costs are charged to the primary product from which gold is recovered. Unfortunately the inexorable increase in costs seems to be the one known and indisputable factor about the future.

6. Conclusions

In conclusion it would, I believe, be useful to summarize some of the factors which will affect the gold mining industry in the medium to long term future. Among the problem areas which will have to be faced are the following:

- a volatile gold price;
- continuing high rates of inflation with consequent effects on operating and capital costs;
- shortages of skilled labour;
- a constantly changing and developing industrial relations scenario.

Despite these rather negative factors there are a host of favourable aspects which endorse the gold mining industry's optimism about the future. These include:

- a broadening gold market with highly favourable implications for the gold price;
- improving productivity particularly in South Africa with the stabilization of the labour force;
- an innovative industry;
- a metal which has stood the test of time and has retained its uniqueness and attraction.

CHAPTER 11

WORLD MONETARY DEMAND FOR GOLD: KEY TO THE GOLD PRICE OUTLOOK

HERBERT J. COYNE

1. Introduction

One might say that gold has often exerted an influence vastly greater than its mere visible presence might suggest. It is a little startling to realize that the sum total of all the gold ever mined on this planet would fill only one-tenth of St. Peter's Basilica. Virtually all of this gold — on the order of 2.8 billion ounces — still exists, with nearly half of it held by central banks and other official agencies, and much of the remainder about evenly divided between private investment hoards and global holdings of jewellery. By far the greater part of this indestructible stock has been acquired as a store of value by private investors and central banks. And changes in their desire to hold gold — which may be called shifts in "monetary demand" — are typically the main force driving gold prices up or down.

In my remarks today, I shall attempt to describe world "monetary demand" for gold, the factors which shape this demand, its global evolution, and crucial importance to the gold price outlook. Over the last few years, world monetary demand has receded moderately, dampened by high gold prices, tight monetary policy and, above all, the competitive appeal of high real interest rates on gilt-edged instruments. It is my belief that this drop-off in monetary demand is a transient phenomenon primarily of a cyclical nature, one that will be reversed before long. I expect that in the next few years monetary demand for gold will resume the growth trend of the past decade, with a buoyant effect on the world gold market and a significant impact on international monetary arrangements. In elaborating these important themes, my discussion today will encompass: 1) the dramatic changes in recent years in the official sphere — in purchases and sales of gold by the world's central banks, and 2) the more significant developments in the private sector, including the successful emergence of new gold markets, the global expansion of marketing and trading facilities, and the increasing variety of gold media available to meet investor interest.

Before I proceed with this rather ambitious survey, let me sketch for you briefly the major cycles and trends in monetary demand which have developed in recent years, and the principal influences which I believe will be uplifting in the years ahead.

2. Key Factors Influencing Investor Demand

Let me begin with a rather broad-brush description of the forces influencing monetary demand for gold (Table 1 and Figures 1 and 2). Alternatively, we can refer to this as "portfolio demand" for gold by investors and central banks. One key influence affecting this "monetary" or "portfolio" demand in the "real price" of gold — especially whether this metal is perceived to be very high or low relative to prices in general. I might recall that, in the late 1960's and early 1970's, a great surge in investor demand developed partly because the real price of gold had been kept artificially low. On the other hand, in 1980-81, an unusually elevated gold price was one of the factors which impaired monetary demand at the same time that it weakened industrial use. An important factor in the current situation is that the gold price has come down to more modest levels, and this is one of the factors auspicious to the renewed growth which I foresee in the coming years — both in official and private demand.

While influenced by relative price considerations, the principal reason people buy gold is as a kind of insurance medium — a hedge against currency depreciation and weakness in other assets. Plainly, the impulse to acquire hedge assets increases when inflation and other anxiety factors are

Table 1 - *World Gold Investment* (million troy ounces)

	1974	*1975*	*1976*	*1977*	*1978*	*1979*	*1980*	*E1981*
Official Coins	9.2	8.1	5.9	4.6	9.2	9.3	5.8	7.5
Bullion	17.8	4.3	—3.8	4.7	1.4	7.2	8.7	4.1
Small Bars	—0.9	—	5.6	2.2	3.3	5.1	0.1	2.0
Medallions	0.2	0.7	1.6	1.6	1.6	1.1	0.5	1.0
Total	26.3	13.1	9.3	13.1	15.5	22.7	15.1	14.6
Value ($ Bil.)	4.1	2.1	1.2	2.0	3.0	6.9	9.2	6.7

E — estimated.

Sources: Consolidated Gold Fields, J. Aron Precious Metals Research Department.

particularly intense. At J. Aron, we have used the expression "World Anxiety Coefficient" to encompass those factors which induce asset holders to increase the gold portion of their portfolios. Now we know that, in the last two years, extremely tight monetary policies have evoked high rates of interest, and have, to some extent, succeeded in moderating the rate of inflation. Unfortunately, we are not sure whether this monetary squeeze can be sustained much longer without a very severe cost to business and employment — perhaps more painful than any government can tolerate. We also know that there are now increasing pressures for reflationary policies in many countries, that high budget deficits are threatening Reaganomics in America, and that the world is still politically schismatic, monetarily unsettled, and faced by an intensified armaments race. In these circumstances, I very much doubt that the World Anxiety Coefficient will ease enough (if at all) to offset other buoyant influences affecting gold demand.

Aside from the influences I've just mentioned, there is a third set of factors which affect monetary demand for gold, encompassing the legal, institutional, and cultural framework within which this metal is sold, the

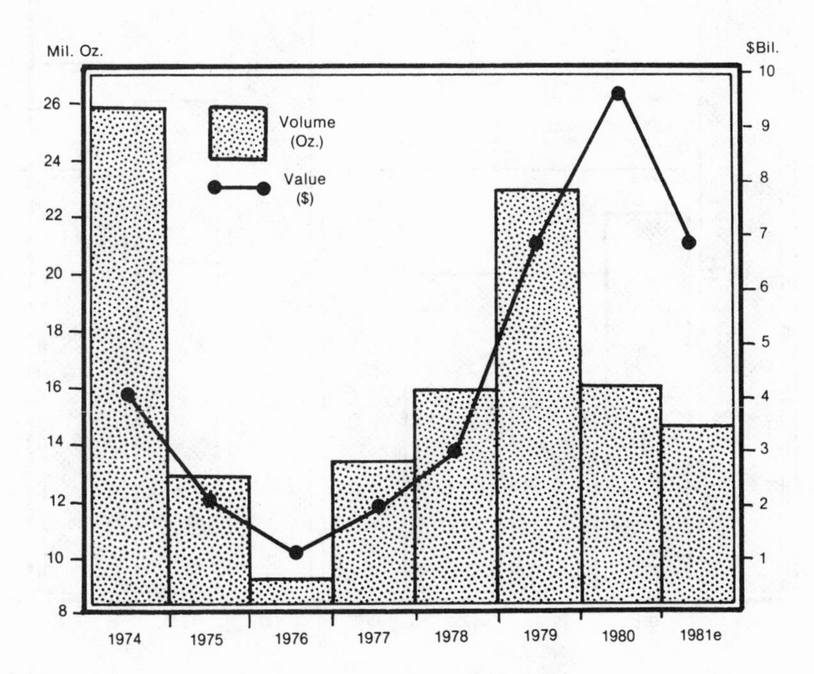

Fig. 1 The volume and value of gold investment (annual data).

e — estimated.

related public attitudes towards gold as an investment medium, and the development of gold's marketing and promotional infrastructure. I think it is highly important to note that, in all these areas, the trends have been very favourable to gold in the last decade and I believe they will continue to be uplifting to the gold market in the years ahead.

Let me proceed now to outline significant developments in the different spheres I have just mentioned: 1) the official sphere of central bank purchases and sales, 2) trends in leading countries and regions, and 3) the new types of gold marketing instruments that are available. Of course, I cannot hope to be all inclusive in my description of this worldwide gold

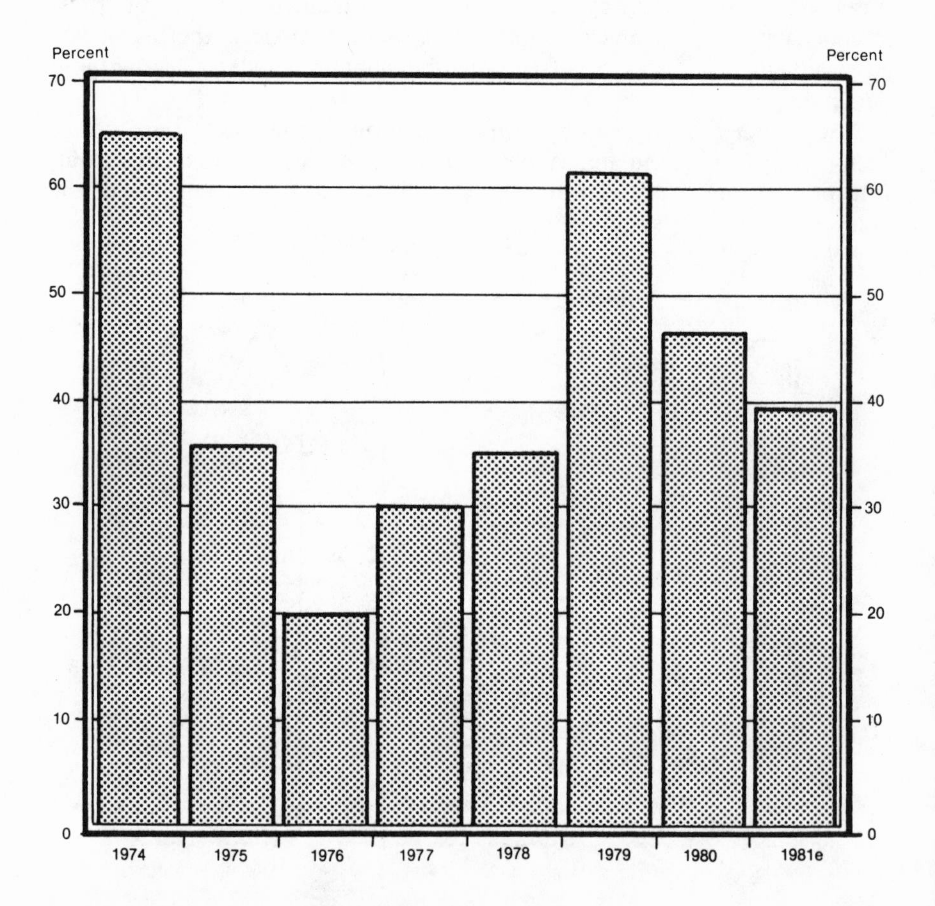

Fig. 2 Investment as a share of total supply (annual data).
e — estimated.
Source: J. Aron Precious Metals Research Department.

investment universe — only to give you a glimpse of developments I regard as especially interesting or portentous.

3. Central Bank Purchases and Sales

Since World War II central banks and other government entities have, on balance, been net buyers of gold (Figures 3 and 4). In fact, on a net basis, the official sphere has absorbed upwards of 150 million ounces over the past three decades. Moreover, in that same period, central banks have been net buyers of gold in 2 out of 3 years.

That buying orientation on the part of central banks was interrupted in a particularly sharp way in the mid-1970's when the IMF and the US Treasury commenced heavy gold sales, as part of an effort to progressively scale down

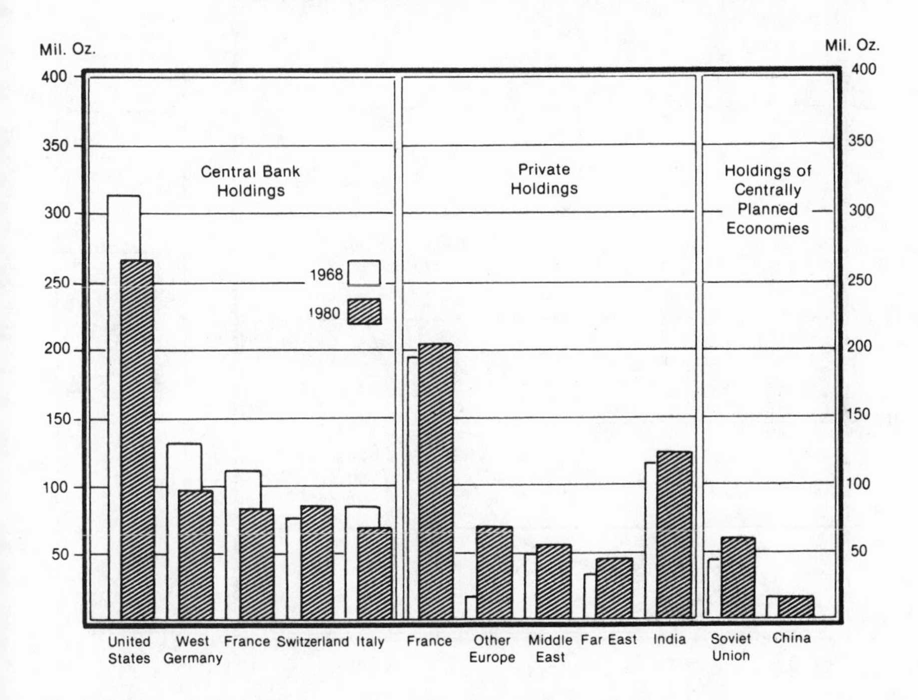

Fig. 3 Gold stocks in selected countries and regions (end of period).

Note: West German, French, and Italian central bank statistics for 1981 exclude amounts transferred to European Monetary Cooperation Fund accounts in 1979.

Sources: International Monetary Fund, Charter Consolidated, Central Intelligence Agency, J. Aron Precious Metals Research Department.

gold's monetary significance. For a time, it seemed to many observers that gold would be yoked indefinitely by the continued movement of that metal out of vast official stocks. And, indeed, very large sales by the IMF and Treasury — coupled with a weaker economic environment — did drive down gold prices in 1975-76. At first this price factor, combined with dollar weakness and oil-fired inflation, contributed to a rise in both private and official interest, especially on the part of central banks and government entities in the Middle East and in many developing countries. Purchases by central banks climbed to more than 10 million ounces in 1980 and then receded (in step with private investor interest) to less than 5 million ounces in 1981. Plainly, government demand for gold, like private investor interest,

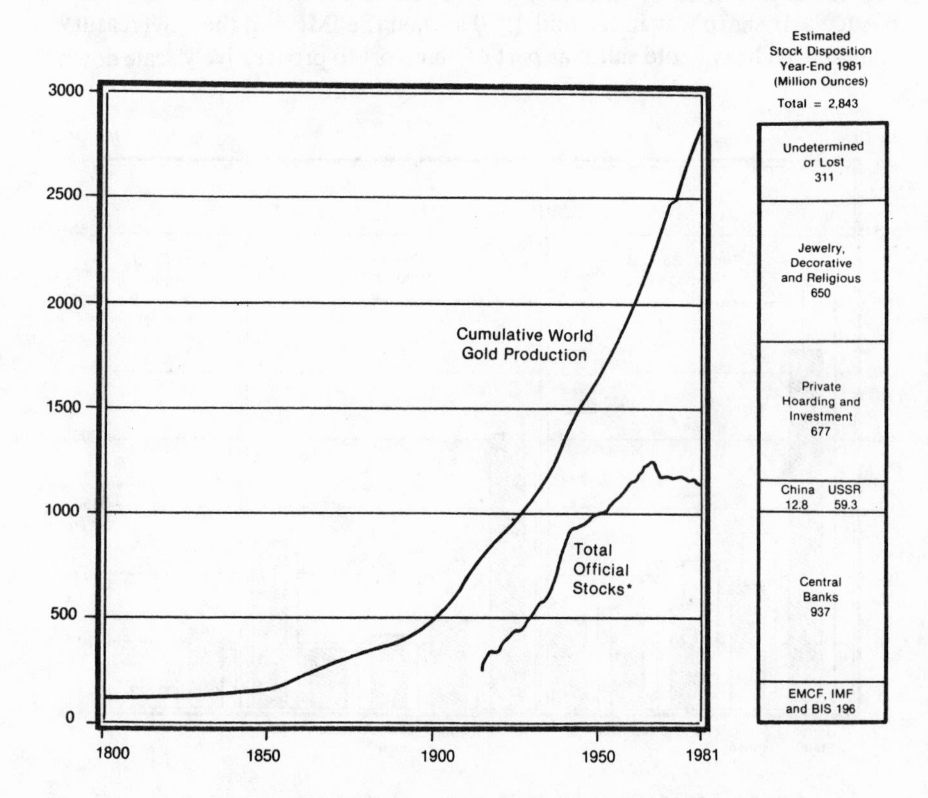

Fig. 4 Cumulative world gold production and its distribution (million troy ounces).

* Including holdings of central banks, the International Monetary Fund, European Monetary Cooperation Fund, and the Bank for International Settlements.

Sources: US Bureau of Mines, International Monetary Fund, Charter Consolidated, Central Intelligence Agency, J. Aron Precious Metals Research Department.

varies in response to changing political-economic conditions.

Nevertheless, there are now persuasive reasons to believe that the longer term pattern of net buying by the official sphere has reasserted itself, and will endure in the future. My confidence rests on several principal developments:

1) The effort to demonetize gold has been blocked or reversed; US Gold Commission deliberations indicate that the period of US Treasury auctions is over and, by implication, the era of large scale official selling is at an end.
2) Interest in gold as a monetary medium has begun to revive — witness its use as backing for the European Currency Unit; also, the increasing utilization of this metal by private banks as an international financing medium will contribute to its monetary acceptance by the official sphere.
3) In a still unsettled world, the impulse to asset diversification is alive and well, and many smaller central banks are conscious of the relative scarsity of gold in their portfolio.
4) We can perceive a mutual reinforcement process by which the increased recognition and use of gold in the private sector abets official interest, and vice versa.
5) In general, a perception of serious disorder in international monetary affairs is a stimulus to both official and private demand for gold.

4. The Growth of the Private Investor Market

Private investors — individual and institutional — have played a markedly increased role in the world gold market since 1965. Despite sharp year-to-year variations in this demand, it appears that fluctuations are taking place around a rising long term trend. This uptrend has been nourished by political-economic factors, and by the worldwide expansion of the "market infrastructure" — the facilities through which gold is distributed and promoted, and the changes in laws, traditions, and attitudes conducive to gold purchases. Also fostering this growth trend in gold demand has been a tendency for major new markets to develop in succession, with an increased flow to one new gold market often offsetting an ebb in an older market.

5. Successive Emergence of New Markets

The door to the US investor market was opened by legalization of gold ownership in 1974-75. This legalization proceeded in two steps: first, removal of the barrier to the import and sale of principal types of gold bullion coin in January 1974; and second, the elimination of legal restraints on bullion ownership a year later in January 1975. The immediate response of the US public was modest — disappointing to those who had anticipated a major gold rush. The development of the US market was initially hampered by the lack of gold consciousness here, the adverse attitude of the media, a somewhat hostile attitude on the part of the government and financial community, the fledgling state of marketing facilities, and the remaining legal and traditional barriers to gold purchases by institutional investors. Moreover, the economic climate soon after legalization proved recessionary and inhospitable to nascent gold demand.

To a remarkable extent, these negative factors have been reversed in the last few years. As I will discuss in more detail later, there has been a highly significant expansion in the gold market infrastructure. In addition, hostility to gold has faded; the media have become much more objective, or favourable in their treatment of gold. In every quarter, we now see gold's acceptance as a reputable investment asset, and a wider appreciation of its usefulness as a medium for portfolio diversification. Gold's appeal is now increasing gradually to banks and financial institutions as well as to individuals. I would estimate that, allowing for year-to-year variations and inclusive of all types of gold, the US investor market is now at a threshold level of around 4 or 5 million ounces per annum; and this level will prove to be a springboard for further gains in the years ahead, subject to limitations of supply availability and price.

In the history of gold market development, perhaps no episode has been more dramatic than the rise of Middle East buying, in the period which followed the initiation of IMF and Treasury sales. It should be noted that the Middle East has been a major investment centre for gold for a very long time. Until 1975, the bulk of this area's demand was in the form of purchases of low mark-up jewellery by a very wide public. The feature of the new gold demand, beginning in the mid-1970's, was that large investors, traders, and government entities, began to absorb larger and larger quantities of gold bullion. I believe that at the outset, these buyers recognized the attractive value which gold presented at that time. As you know, this demand was also greatly buoyed by the unprecedented flow of wealth to the Middle East, with the annual OPEC surplus in some years soaring to well over 100 billion dollars.

At a time of peak international tension and dollar weakness, gold purchases by large private and official buyers in the Middle East climbed to an estimated annual rate of over 8 million ounces per annum. This demand has subsided in the last few years, weakened by the same factors which have been affecting investor interest elsewhere, and also by a decline in the OPEC surplus. Nevertheless, the stock of accumulated wealth and the flow of funds into the area remains enormous; and the practice of diversifying portfolios into gold has become well entrenched. As in the US, we have seen the development of a solid base for future growth.

In this very recent period of subdued world investor demand — there has been one singular bright spot in the gold picture — the rising sun of Japan and other Far Eastern markets. J. Aron's Research Department estimates that Japan imported about 5.5 million ounces in 1981, up more than 50 percent from its level a year earlier. It appears that nearly one-half of this import volume was acquired by investors. The impetus to this surge in Japanese demand came, in part, from new tax considerations associated with the establishment of a "green card system" enabling the government to closely monitor private investment returns. Aside from these tax considerations, gold has also benefitted from the removal of restrictions on gold ownership, from an increasing gold consciousness in Japan, and from the rapid development of gold marketing facilities including the establishment of an Intergold office in Tokyo in late 1980, the initiation of bank and brokerage house offerings of gold in the form of coins, small bars and gold certificates, and the establishment of three private gold exchanges in 1981. A major event early this year will be the opening of a gold futures exchange in Tokyo. It seems to me highly probable that Japan will become one of the world's leading gold markets, and perhaps the foremost investor outlet for gold.

My brief comments on developments in each of these countries is simply intended to point up a tendency for the successive emergence of new markets to foster overall growth in the world investor market. It is plain that in all of these countries growth in the "market infrastructure" has been instrumental in bringing investor demand to its present level, and creating potential for future growth.

6. The Development of the Market Infrastructure

Over the past decade, net investor demand has fluctuated widely around a long term growth trend. One principal impetus to that growth trend has come from environmental factors — especially from stubborn world inflation. Growth in investor demand for gold also has been greatly nourished by the progressive development of global marketing and promotional facilities, the breakdown of legal and institutional barriers to holding gold, and an associated growth in awareness of gold as a legitimate portfolio diversification medium. This expansion in the gold investment infrastructure continued to be very much in evidence in 1980-81, though with an uneven impact. Some investment sectors forged ahead while others consolidated or waned.

For the most part, weakness in some gold investment sectors has been prompted by exceptionally high gold prices or by changes in the political and economic environment (e.g., high interest rates, a stronger dollar). These influences have masked the continued development of the underlying gold marketing and institutional base. Some highlights of this development in 1980-81 are as follows:

— In the US, prominent banking and investment institutions began to directly service clients in gold and also to market gold to the public. The development of gold deposit accounts at major banks has accelerated. Leading banks and brokerage houses are actively retailing gold to consumers, either through gold passbook accounts or via gold certificates. These gold certificates are now offered around the world by various financial institutions. Innovative programs, including bullion and coin purchasing services and gold "share-builder" accounts, have been established.

— A growing variety of bullion coins have become available to investors and traditional suppliers have continued their aggressive marketing programs. Bullion coins have long been popular among individual investors because of their high liquidity. The more widely known coins, such as Krugerrands, Maple Leafs, and the Gold Coins of Mexico are easily sold in large or small quantities with assay at full fair price for immediate cash payment. As a consequence, wide public acceptance of these coins now provides a foundation for the global expansion of this market. In 1980, the introduction of bullion coins in quarter and half-ounce denominations has helped to maintain gold's affordability to "small" investors.

— Some principal banks have initiated their own gold bullion trading departments. This may well encourage the utilization of gold by these banks in their regular investment and portfolio management services, as it has in West Germany and Switzerland. Bank and brokerage firms are also offering gold bullion widely in the form of small bars. Sales of these small bars have been especially popular in the rapidly developing Far Eastern markets. In Japan, several leading trading houses have been actively advertising and promoting small bar sales.

— It is highly significant that some pension funds such as the Alaska State Pension Fund have begun buying gold, while others are seeking the necessary approval of their governing boards before initiating purchases. In 1979, regulatory changes in the US proved to be a milestone in eliminating legal barriers to gold investments by pension funds. These funds are among the largest asset managers in the United States, accounting for more than 20 percent of investable funds in the trillion dollar range. Thus, a money-manager market of enormous potential has begun to open up to gold and, by example, to encourage gold purchases by other fiduciaries.

— With new markets opening in the Far East and elsewhere, gold trading operations have been spreading around the world. This metal has virtually become a "24-hour market". And the introduction of gold options trading on organized exchanges is further enhancing the ability to transfer risk and increasing the diversity of gold investment media available to investors. In general, gold trading has been expanding globally, in line with the development of the gold market. In turn, the growth of gold trading activities helps to stimulate the development of gold marketing and promotional facilities.

7. The Increasing Diversity of Gold Media

The past decade has seen a remarkable multiplication in the variety of gold media by which individuals and institutions can participate in the gold market.

Ten or fifteen years ago, gold availability to investors was pretty much limited to physical metal — mostly bullion, small bars, and coins and, in some countries low mark-up gold jewellery. Gold coins, today the single most popular gold investment vehicle (Figure 5) — were not aggressively marketed or widely available to investors 15 years ago. (In the 1960's, bullion coins accounted for less than 5 percent of annual gold consumption.) These coins began to be significant investment vehicles in the mid-1970's; today they consume an unusually large share — between 20 and 30 percent — of newly mined gold entering the market each year.

In analyzing the phenomenal growth in sales of Krugerrands and other gold coins over the past decade, it would be hard to over-emphasize the significance of the superb distribution and marketing system established by the South African Chamber of Mines. The success of the International Gold Corporation, the marketing subsidiary of the Chamber of Mines, is an inspiration to others who are actively marketing gold and introducing new gold media.

Coin, bullion, small bars, and medallions — the four traditional means of physical investment in gold — are still the major media employed. J. Aron's Research Department estimates that, of the 130 million ounces of physical gold added to investor holdings over the eight year period 1974 through 1981, coinage accounted for the largest share of the total. About 60 million ounces of gold coin were added to investment stocks during this period, accounting for 46 percent of the increase in total investor stocks (excluding jewellery). Bullion is the next largest sector. In 1974-81, investors acquired 44 million ounces in this form, 34 percent of their total offtake. That leaves about 26 million ounces (or 20 percent) which was purchased in other forms, namely small bars or medallions.

It is interesting to contrast the behaviour of these different gold media. Bullion demand appears to be the most volatile component. J. Aron estimates that world demand for bullion has fluctuated from net purchases

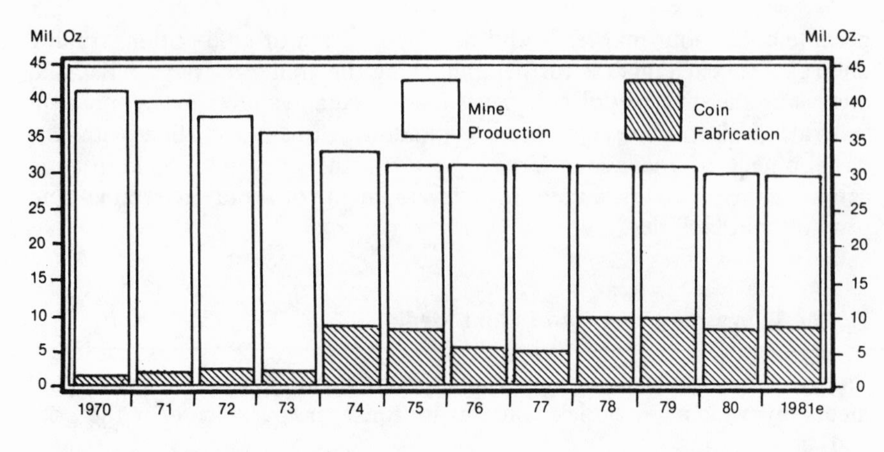

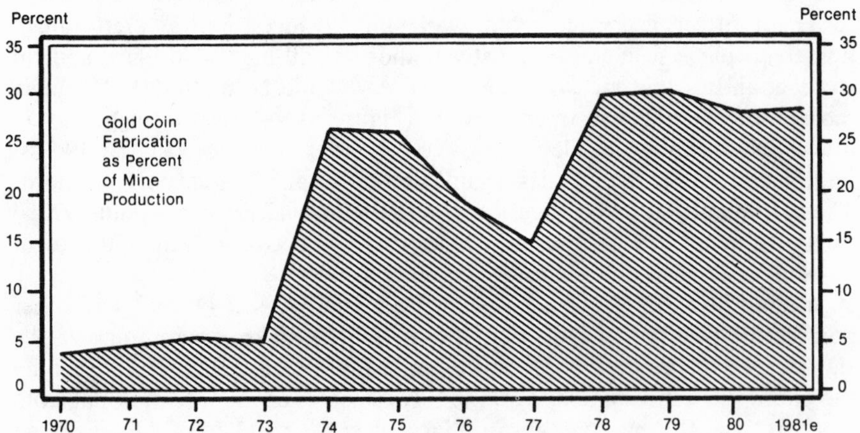

Fig. 5 World coin fabrication vs. mine production (annual data).

Note: Excluding centrally planned economies.

Sources: Chamber of Mines of South Africa, Statistics Canada, US Bureau of Mines, Consolidated Gold Fields, J. Aron Precious Metals Research Department. 1981 estimates: J. Aron.

of 17 million ounces in 1974 to net dishoarding of nearly 4 million ounces in 1976, with a rebound to around 8.9 million ounces per annum in 1979-80. On the other hand, while demand for coin fluctuates, it has never become a net source of dishoarding or "negative demand" in any given year on a worldwide basis. Obviously some coins return to the market as scrap or are resold into the investment market in any given period; nonetheless, the gold coin market has been relatively stable in recent years, fluctuating in a range of 4.6 to 9.3 million ounces per year.

In the past few years new types of investment have been introduced, including precious metal certificates of deposit, investment accounts at banks and brokerage houses, leverage accounts, and mutual funds that concentrate on gold and other precious metals. Looking at these media, we can see that to some extent they are patterned on similar instruments in financial markets. More recently these media have been marketed in the Far East and South East Asia. It is important to note that these new and sophisticated gold media have behind them the physical reality of either bullion or coin, so that in effect they are extensions of the traditional gold investment media. Sometimes the imposition of value-added taxes, sales taxes, and other restraints to gold investing have encouraged the introduction of these paper instruments in lieu of the physical bullion or coin.

Hand in hand with the growing menu of gold investment possibilities has been the emergence of new and diverse players, or groups of investors, turning to gold. The past decade and a half has seen investor interest spread from a relatively small number of wealthy individuals (except in certain countries like France and India which have long traditions of gold ownership) to the middle class in Europe, America, and Japan. Indeed, after centuries of being the investment of kings and national treasuries, gold over the past century has been in the process of becoming increasingly accessible to everyone. Thus gold has been experiencing a process of progressive "democratization", which will ultimately contribute further to the expansion of the world investor market.

Since we have covered a lot of territory in realms of gold, it may be helpful for me to briefly recall some highlights of our voyage:

1) As a rule, variations in world "monetary demand" — in the desire to hold gold as a store-of-value medium — have been large enough to outweigh year-to-year changes in ordinary gold supply/demand categories. Accordingly, shifts in monetary demand have usually dominated price movements in gold.

2) Although monetary demand has receded in the last two years, this decline is a temporary, and very largely cyclical, phenomenon.

3) Growth will very likely resume in the years ahead, because of
   a) the persistence of global tensions and anxieties conducive to portfolio demand;
   b) the strong desire for an official reserve medium that is apolitical — and not a liability of any government entity;
   c) the regular emergence of new markets (Japan is the latest example);
   d) the development of the gold marketing and promotional infrastructure; and the progressive elimination of legal and institutional barriers to gold demand;

e) the ready availability of a broad, increasingly varied "menu" of alternative gold products to suit a variety of tastes.

All of these factors, needless to say, are uplifting to gold, and they buttress my confidence in this metal's long term outlook. In my view, the recent dispirited state of the gold market is no indication at all of this metal's prospects, any more than an interlude of despondency when IMF/Treasury sales began had any enduring significance. As I see it, gold is nearing a cyclical low. It is my belief that 1982 will prove, in retrospect, to have been an outstanding gold buying opportunity, and the starting point of a new upturn in "monetary demand" by central banks and investors.

CHAPTER 12

GOLD: FUNDAMENTAL INFLUENCES UPON THE INVESTMENT CLIMATE*

ROGER C. VAN TASSEL

1. Forecast: Sizeable Uncertainty

During the 1970's gold made an impressive record as an investment. It outperformed the returns on stocks and bonds; it provided a rate of return that substantially exceeded the rate of inflation. It retained an important role as an industrial commodity and, in spite of continued steps toward demonetization (the official closing of the gold window by the United States in 1971 and the official gold sales by the IMF) we are witnessing serious discussion of the desirability of the restoration of gold to a position of renewed influence in the international monetary system.

Forecasts of the price of gold or its future monetary role, as is the case with forecasts generally, must contain sizeable elements of uncertainty. Events do occur affecting the course of economics that cannot readily be included in forecasting models. However, gold has a role both as an investment media and as a commodity. It is a mineral produced under conditions that are very well known. These commodity aspects have had fundamental influences upon the price of gold in the past, have tended to restrain price oscillation created by its investment role, and offer a degree of comfort for economists seeking insight into future trends. The major influences upon the price of gold over the last decade have been identified. Their continuation, or replacement by equally powerful forces, is required if gold is to continue its powerful record as in the 1980's.

---

* Part of this article is based on continuing reasearch supported by the J. Aron Co. Demetrius Kantarelis was of great help as a research assistant in completing this paper.

2. The Investment Record of Gold

The price performance of gold during the decade of the 1970's was impressive. Table 1 compares the price change of gold against the (CPI) Consumer Price Index in the United States, against selected commodities and commodity indexes, and against the price changes of selected stock and bond indexes. The mean return on gold was substantially better than alternative classes of investments. Obviously, investment in a particular well selected stock could have been better; purchase of investment grade diamonds at the correct time would have performed substantially better; but the overall performance was much better than most readily marketable, alternative types of investment.

Table 1 - *Rate of return* (average per month in selected time periods[1])

	1/69-12/77	*1/69-1/80*	*1/69-5/81*
Metals:			
Gold	1.234	2.102	1.690
Price Indexes:			
Thirteen Raw Industrials	.670	.687	.716
Crude Materials for Further Processing	.689	.801	.815
Nonferrous Metal	.578	.858	.699
CPI	.520	.592	.642
Bonds:			
Moody's AAA Rates	.204	.395	.540
Commercial Paper	.021	.502	.651
Stocks:			
Standard and Poor	—.079	.063	.177

Sources: *Minerals Yearbook*, US Department of the Interior, various issues (for the gold prices); *Survey of Current Business*, US Department of Commerce, Bureau of Economic Analysis, various issues (for the price indexes, bonds and stocks).

The volatile price pattern in 1978, 1979 and 1980 substantially affected the average return per month. January 1969 to December 1977 shows an average monthly gain of 1.23 percent; extending the period to the $850 peak price in January 1980, gives an average monthly gain of 2.01 percent; and continuing through the more recent price decline to May 1981 shows an average gain per

[1] Given two sets of data for a function — price of gold — over time, each can be expressed in terms of: $S = Pe^{rt}$. For example, if sales equal 4.36 in period 1 and 6.87 in period 4, substitution in the general formula gives: $4.36 = Pe^{r(1)}$; $6.87 = Pe^{r(4)}$. Taking the natural logs of each equation: $\ln 4.36 = \ln P + r$; $\ln 6.87 = \ln P + 4r$; or $1.47247 = \ln P + r$ (1); $1.92716 = \ln P + 4r$ (2).

month over the whole period of 1.69 percent per month. Gold was not a certain winner, however, even during this period. This is especially true for the short term investor. As noted by David Potts, "Gold was in a bear (falling) market during most of 1980 and the beginning of 1981. The extent of the fall from the peak price of US $850 per troy ounce on the 21st January, 1980 is comparable to that which occurred in the mid-1970's when the price of gold dropped from US $197 to US $103 over a period of 20 months."[2]

These large price shifts have largely resulted from shifting investor attitudes. Nevertheless, these price changes have exerted a strong influence upon purchase of gold for use in fabrication.

In both major declines — if we can assume approximately US $400 represents the bottom of the present bear market — gold has fallen to roughly 60 percent of the previous peak price. Trend analysts take comfort in this information and predict a resurgence in the price of gold in the near future. This may happen; but examination of the forces that gave gold its previous two boom periods introduces an element of caution.

Gold in 1975-76 and again in 1980-81 suffered substantial price decline. Silver also showed negative returns in 1975 and again in 1980. After the massive increase from $6 per ounce at the start of 1979 to a peak price of $45 in January 1980, silver fell all the way to $12 by late May of 1980. There were periods of great short term gain; over a longer period the precious metals did better than most alternatives; however, the wide price savings demonstrate very large losses possible for investors buying at or near peak prices.

Table 1 shows another dimension of gold's value in an investment portfolio. Not only did gold present a higher return than available in stocks, but the price movements in gold correlated negatively with stock prices giving the opportunity for both greater return and reduced risk.

The negative correlation for gold stocks with other stocks was much stronger than the relation between gold and stocks generally. This was noted by Mr. John van Eck in a speech before the New York Society of Security Analysts on March 8, 1977. "From the first quarter of 1968 when International Investors began to concentrate on gold mining shares, through the fourth quarter of 1976 the S and P 400 had an average annual rate of return of 5.6 percent and a risk of 9.6 percent, the risk of being the variance above or below the mean return. I.I.I. had an average annual return during this period of 6.9 percent with a risk of 19.7."[3]

(continued)

Subtracting (2) from (1) to eliminate ln p, $-0.45469 = -3r$ or $r = 0.15156$. Therefore, the return rate per period is 15.156 percent.

[2] David Potts, *Gold 1981*, Consolidated Gold Fields Ltd., May 1981, p. 7.

[3] John C. van Eck, "Investment Policy Strategies", *Wall Street Transcript*, 29 March, 1977.

Even though gold mining shares were volatile they did reduce overall portfolio risk because they tended to move in the opposite direction from industrial shares. Similar evidence was represented by McDonald and Solnick tracing the correlation between the prices of "gold" stocks and the Standard and Poor's 500 stock index for a 28 year period betweeen 1948 and 1975. They found a correlation of -.40.[4]

3. Gold's Commodity Role and Long Term Price Trend

The fact that gold has an important commodity role is central to explaining why the long term price for gold should rise in real terms and why speculative buying or selling of gold tends to be self-correcting. The price of gold may once again show sharp oscillations. But barring a successful restoration of an official fixed price gold system, the long range price of gold will follow a path consistent with balancing the flow of new gold with the demand for industrial use plus any long term purchases for private or official holdings.

Reviewing recent price changes the J. Aron Company noted how the connection between commodity and investment markets serves to limit price movements. "Throughout this time of gold depreciation, however, the market's balance wheel continues to function. Just as rising prices in earlier years had caused a retrenchment of industrial gold requirements, the much reduced price levels of 1976 and 1977 fostered a recovery of fabrication demand."[5]

The gradual collapse of the Bretton Woods system has many explanations: failure to control domestic inflation in major nations, an inadequate adjustment mechanism, and increased use of increasingly illiquid dollar balances. In a sense, however, we could also say the failure of gold mining to produce adequate additional reserves coupled with a weakened adjustment mechanism led to the increased use of dollars and eventual formal suspension of convertibility on August 15, 1971. Among the reasons for the inadequate growth of reserves was the fixed price of gold with its effect upon the production and sale of gold and upon the growth in the industrial uses of gold.

By the end of the 1960's it was becoming increasingly apparent that the $35 price (much too high given gold's value as a commodity in 1934) was becoming much too low. Instead of gold flowing into official reserves, increasingly gold left official reserves to meet growing industrial uses. There

---

[4] John G. McDonald and Bruno H. Solnick, "Valuation and Strategy for Gold Stock", *Journal of Portfolio Management,* Spring, 1977.

[5] J. Aron Co., *Gold Statistics and Analysis*, Dec. 1981/Jan. 1982, p. 78

was a growing awareness by investors that the official price was too low in real terms and would not last. With the advantage of hindsight of 15 years, perhaps our biggest surprise should be that the rise in the price of gold was not even more rapid.

How does the commodity role affect the investment role? Why should gold serve as an effective hedge against inflation? A few assumptions and estimates about the commodity market for gold help us focus on these questions. The assumptions are as follows: "(1) the quantity of gold available for private purchase will remain in the vicinity of 1600 tons per year, (2) income elasticity of demand for gold is between +1.2 and +1.5 in advanced nations, (3) price elasticity is between -.33 and -.7 for gold used in industrial applications in advanced nations, and (4) real income will rise by 3 percent per year in advanced nations."[6] These assumptions combine to produce a projected real price increasing from between 6.4 and 13.6 percent per year. This range, of course, relates to a long term trend. If the flow of new gold slows, as it appears to, the price should rise more rapidly. Certainly we should not anticipate sales from the IMF and national reserves. If long term investment demand weakens, the price of gold should move up more slowly. If industrial output grows more slowly or if price elasticity of demand increases, there would be a slower increase in price.

Predicting price changes, given gold's commodity and investment roles, is complicated by the peculiar stock-flow relationship. Even without any net investment demand - private or official - the commercial demand would begin to press upon new output at a constant real price. New production, even at higher real prices, is anticipated to decline.[7] However, the ratio of gold in public and private stocks to annual use is perhaps 30 or 40 to 1. If large sales from these stocks were to occur, the effect on price would be dramatic. But at this time sales from official reserves do not seem likely even without the question of official restoration of gold to the international monetary system. While restoration should add to official demand for reserves, it could also destroy the private investment market.

Large scale, short term buying of gold, when added to commodity demand and long term investment purchases, can force the price of gold to

---

[6] Roger C. Van Tassel, "The New Gold Rush", *California Management Review,* Winter, 1979, pp. 30-31.

[7] The Chamber of Mines of South Africa published an estimate (Vol. 3, No. 1, Feb. 12, 1980, p.3) indicating little increase in production from a higher real price assumption. "Two gold prices have been used: a supposed current price of $305 an ounce rising to $407 an ounce by 1984 and remaining constant (i.e., the price will rise at the same level as working costs) until the year 2000. The second line projects a price of $450 an ounce rising to $554 in 1984 and then remaining constant until 2000". The total outputs estimated over this period are virtually identical. In fact, until the late 1980's the lower price is estimated to produce slightly larger output. Such a reaction is not unreasonable to expect. See Roger C. Van Tassel, *op. cit.,* p. 30.

rise rapidly well above an equilibrium price. But as the price rises, commodity usage declines. As temporary pressures cause price increases, more gold must be absorbed by the speculative market. The same mechanism works in reverse. If selling forces down the price of gold, industrial usage grows and gradually exerts an upward pressure on price. "In summary, the most fundamental reasons why gold can be regarded as a long term inflation hedge rest upon the supply of gold as an exhaustible resource and in the demand for gold as a commodity."[8]

4. Price of Gold, Inflation and Interest Rates

If one is interested in the probable direction of the price of gold over the balance of the century, the fundamental influences have probably not changed dramatically in the last few years. Perhaps some of the estimated price and income elasticity figures have altered, available new gold may decline more quickly than expected, and possible changes in the use of gold in official currency reserves all may affect the trend. However, the major changes affecting the general economic climate and gold in particular are already visible and will dominate in price movements over the next several years. Inasmuch as gold has typically acted as a sensitive barometer to changes in economic policy, it should not be surprising that expectations of success or failure for the Reagan economic policies dominate much of the thinking in the United States regarding the near future of gold. While there are varied expectations, most forecasts for gold to reach $4,000 within the next several years are based on a fear of hyper-inflation. This expectation — most common among the "far right" — rests upon a belief that the Reagan program to control inflation either will not work or will work only at the cost of a recession sufficient to bring into power another "liberal", high-spending regime. The reaction of the leaders of the Democratic party is similar, except of course they do not see their restoration leading to hyper-inflation. Even though the rate of inflation has been reduced to perhaps half of the previous year's, concern remains that this decline is temporary. If the federal deficit is not reduced more quickly than seems likely, the alternatives seem to be heavy government borrowing in the bond market forcing interest rates up or a return to monetary expansion and renewed inflation.

In a January 1980 report, Peter A. Gilbert[9] presented a chart (Figure 1) that shows the closeness of the relationship between inflation and the price of gold. He notes the dual characteristics of gold. It acts as a normal

---

[8] *Ibid.*, p. 31.

[9] Peter A. Gilbert, "Gold and Related Equities in an Investment Context", *Industry Report*, Wells Fargo Bank, January, 1980, p. 18.

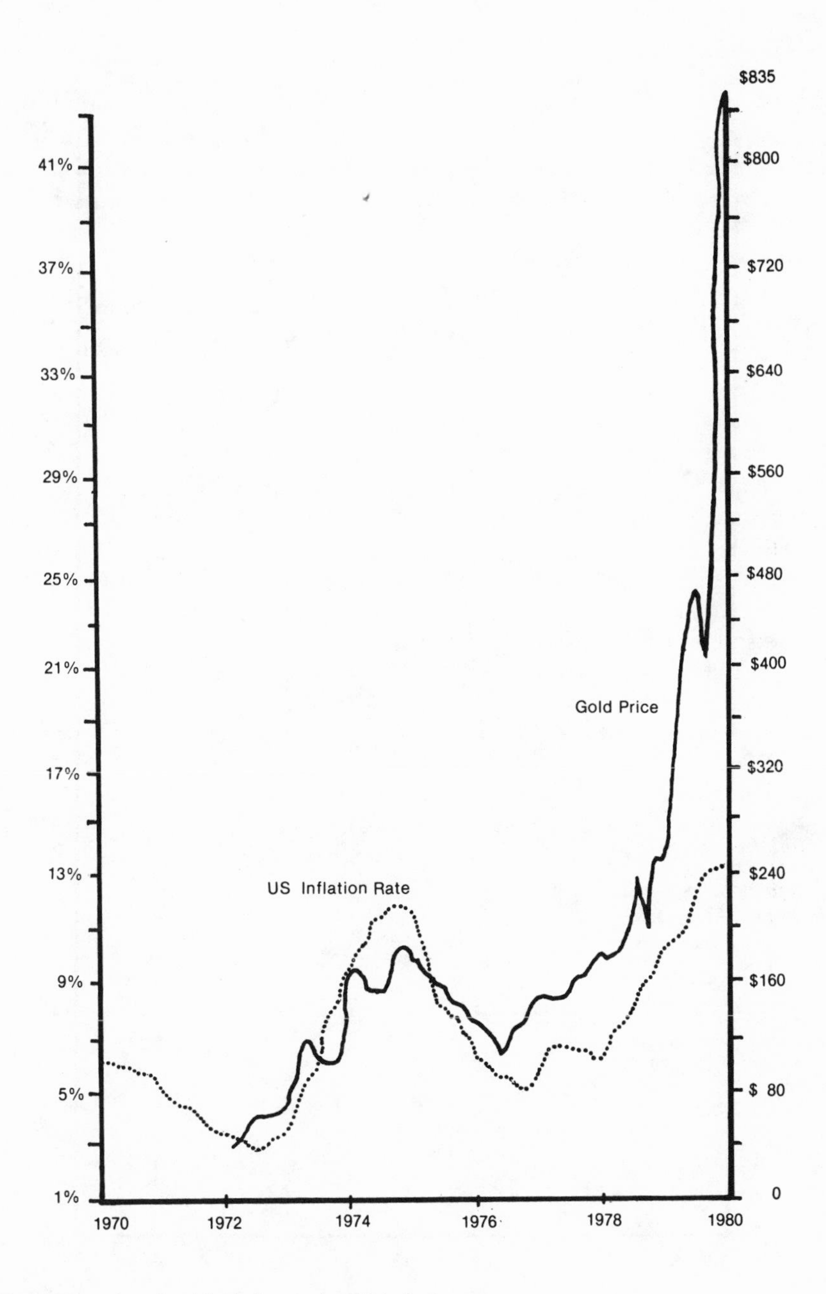

Fig. 1 The price of gold relative to the US inflation rate.

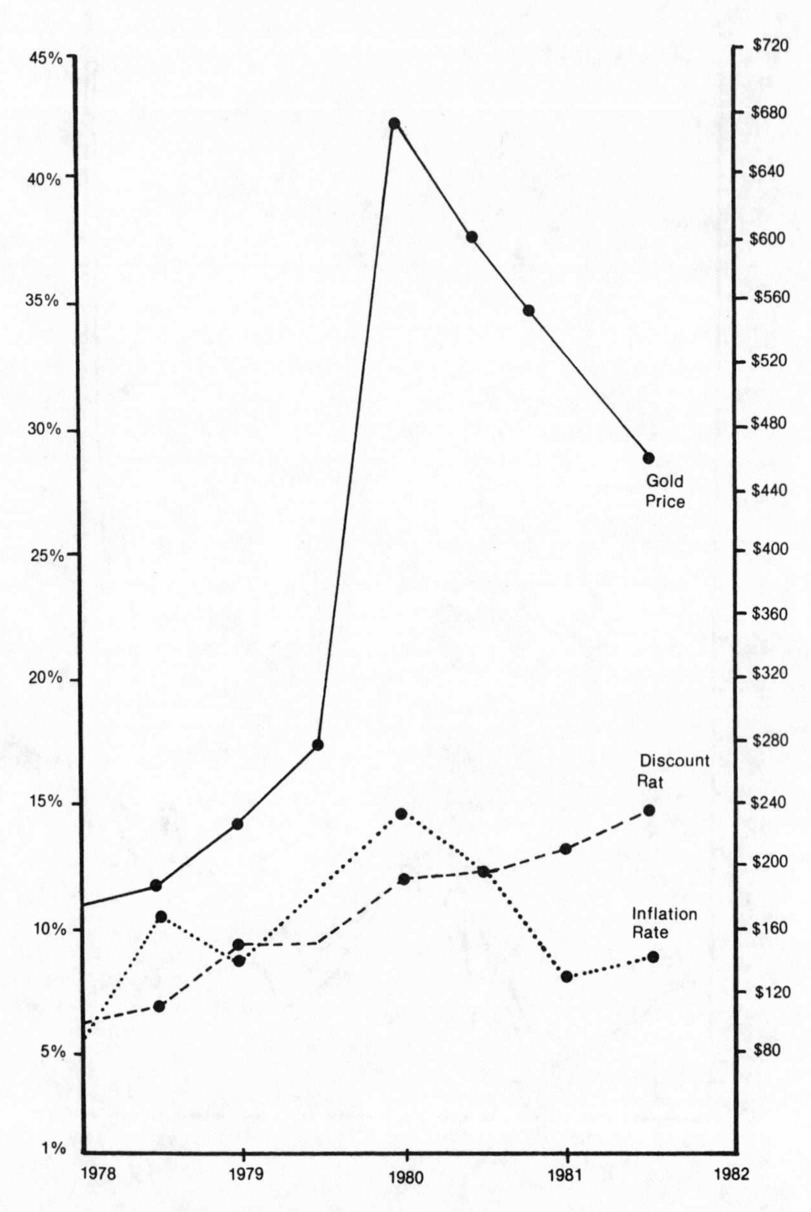

Fig. 2 The price of gold relative to the US inflation rate and discount rate.

commodity with an equilibrium price set by normal supply and demand. But, "additionally, and perhaps more importantly, also displays the characteristics of a financial asset. It is held in anticipation of an investment or speculative return and it is utilized as a hedge against monetary depreciation, i.e., as a result of cautionary demand."[10]

While the price of gold and nominal interest rates have generally reflected the rate of inflation, the price of gold and interest rates move in opposite directions as shown in Figure 2.

If gold can be bought and held at low interest rates, and especially if other familiar investment media are not attractive, gold becomes an attractive haven. In the last months of 1980 and most of 1981 we witnessed the powerful depressing effect on the price of gold of a sharp turn toward positive real rates of return on financial assets. Nevertheless, gold has been relatively stable for several months with the continuing uncertainty about the effectiveness of national policy in keeping the rate of inflation under control. For similar reasons there has not been a shift to long term bonds sufficient to bring down nominal rates dramatically. There remains fear of renewed inflation bringing capital losses and the opportunity of better yields. Equities remain at record low price-earning ratios and have yet to demonstrate real strength. Success for the Reagan program — ending double digit inflation without a deep or protracted recession — would mean boom in equities, bond prices, and decline in the price of gold.

Will the economic chemotherapy kill the disease of inflation without also killing the patient by an overdose of recession?

5. Other Influences on the Role of Gold

Several other influences should be noted. Obviously gold marketing arrangements have been highly developed. Gold still represents a tiny fraction of savings. A very modest shift toward gold would have great impact. For example, the institutional investors in the United States alone control investment funds of approximately $1.5 trillion. Less than one percent of these assets would have been sufficient to have acquired the entire supply of new gold entering the market in 1981 given prices in effect.[11] Many gold newsletters continue to cite this as a bullish influence. However, even should gold attract more institutional investment we should expect the commodity role of gold to exert a balancing influence.

---

[10] *Ibid.*, p. 18.

[11] J. Aron Co., *op. cit.*, p. 64.

In the mid-1970's gold fell by 40 percent and after 20 months rose from $103 in 1976 to a January 1981 peak of $850. We have since experienced another 40 percent decline and more than 20 months of a bear market. If gold resumes its advance it will not be a pattern repeating mechanically but because similar (or new) forces cause the increase. Previous booms have had identifiable causes. So must any future boom. It is not at all sure causes of a future boom can be found over the horizon.

Crisis, of course, can and will erupt. But recent crises have not seemed to have the same power over short term price movements. Change of government, the recent attempt to kill the Pope and President Reagan, and continuing crises in Poland and in the Middle East seem to have had little effect. Gold has advantages as an investment in crises but perhaps people of wealth needing a safe haven for some of their wealth have acquired the needed "insurance" and now judge gold as a more conventional investment medium. Portfolio adjustments have taken place following the gradual removal of gold's role in the international monetary system and its growth as a private investment asset.

The potential restoration of "official" gold remains an influence. While support for restoration of a gold standard is limited, while the type of system that would be attempted is unknown, while the new official price — above or below a commodity equilibrium — is not known, the serious, official consideration of a restored gold standard seems both to provide a degree of protection against downside risk and reduce the opportunity for gain.

In summary, while there are adequate reasons — based on its commodity role — for the long term price of gold to outpace inflation, the balance of forces affecting the price of gold recently have been bearish. While there is mechanism that warrants a continuing long term increase in the real price of gold, it appears the current market in gold and securities seems to be waiting for clearer signs from the Reagan economy. Success for the Reagan program in arresting inflation, especially a resurgent stock and bond market, would devastate the gold market.[12] On the other hand, a return to rapid inflation and easy money would create a rapid upward movement of gold both in nominal and in real terms. At the moment it is uncertain whether the real economic growth in the United States will prove sufficient to permit the reduced deficits essential to many of the central objectives of policy. Without more rapid growth there must be either an abandonment of important domestic policy objectives or reduced military outlays sufficient to enable progress toward a balanced budget.

---

[12] A growing economy affects gold in contradictory ways. More rapid growth adds to the commodity demand for gold. But with increasing returns conventional investments, the investment demand for gold would likely weaken to an extent more than counter balancing the positive effect on industrial demand.

References

J. Aron Co., *Gold Statistics and Analysis,* Dec. 1981/Jan. 1982.

Gilbert, Peter A., "Gold and Related Equities in an Investment Conflict", *Industry Report,* Wells Fargo Bank, January, 1980.

McDonald, John G. and Solnick Bruno H., "Valuation and Strategy for Gold Stocks", the *Journal for Portfolio Mangement,* Spring, 1977.

Potts, David, *Gold 1981*, Consolidated Gold Fields Ltd., May, 1981.

van Eck, John C., "Investment Policy Strategies, *Wall Street Transcript, 28 March 1977.*

The Chamber of Mines of South Africa, Vol. 3, No. 1, February 12, 1980.

Van Tassel, Roger C., "The New Gold Rush", *California Management Review*, Winter, 1979.

CHAPTER 13

ASPECTS OF THE INDUSTRIAL DEMAND AS TO JEWELLERY: CASE STUDY OF A TYPICAL COUNTRY

LUIGI STELLA and FABIO V. TORBOLI

1. The Relevance of the Topic

We wish to give full credit to the organizers of the Conference for having carefully considered the programme and topics to be discussed. This Conference on Gold would certainly not have been complete had it not included the jewellery market, which alone, in the decade from 1971 to 1980, processed over 7,100 tons of newly mined fine gold, equivalent to 52 percent of the gold available for "net private purchases" in that period.

We would also add that the Italian gold jewellery industry represents a very important share of this market. In fact, between 1971 and 1980, the Italian manufacturers processed 1,650 tons of fine gold, equivalent to all the gold mined in South Africa in two and one-half years.

This paper is intendend to give a general picture of the jewellery sector, and a brief review of the Italian jewellery industry, its present state and its prospects for development.

We hope in the short time available to us to make it quite clear that the gold jewellery market is fundamental to the industrial demand for gold; and we are confident that our comments on jewellery will hold the attention and interest of the few ladies attending this meeting.

2. A Survey of the Italian Gold Jewellery Industry: the Biggest in the World

We don't wish to start our speech by going back too far in time, but we must at least mention Italy's first world-famous goldsmiths: the Etruscans, whose skills in embossing, granulation and filigree have been admired and envied through the ages by every generation of goldsmiths.

In the sixth century B.C. the Etruscans were already producing large quantities of gold rings with figurative gold settings. As their reputation grew, and demand for their work increased, they established large and well-organized jewellery workshops, which were the forerunners of today's factories.

The most famous of these workshops — and of the earliest gold jewellery industries in the world — was located in a place called Vetulonia, known nowadays as Poggio alla Guardia near Grosseto.

However, jewellery production on an industrial scale, for which Italy is famous today, is very recent and dates back to the time when the first automatic chain-manufacturing machines were introduced in about 1920; when Giovanni Balestra extended his plant at Bassano del Grappa, and Gori & Zucchi, the largest gold jewellery manufacturer in the world today, were about to enter partnership and set up a factory in Arezzo.

On a nation-wide scale the development of the jewellery industry was slow. First came the long, painful period brought by the 1929 crisis; later, industry suffered from the autarchic regulations of the time and the scarcity of gold, thus causing a slowdown in growth. Then came World War II: factories were destroyed, machinery was stolen or rendered useless owing to the lack of spare parts, and there was little or no gold to make into jewellery.

Postwar reconstruction started slowly in the early 1950's. By 1960, Arezzo had about 30 gold jewellery manufacturers, Vicenza about 130 and Valenza Po had already become famous as a miniature capital of jewellery manufacture with over 500 small local firms catering to every requirement of the trade.

In the following decade, the industry developed rapidly, and by 1970 the number of manufacturers in Arezzo had tripled to reach 90. Vicenza had 280 and Valenza Po 1,000 jewellery manufacturers and artisan producers.

The same period saw the beginning of an unprecendented expansion in Italian gold jewellery exports. In the early 1970's, the industry processed over 160 tons annually. This increased to a record 235 tons in 1978, 27 tons

Table 1 - *Italian gold jewellery industry* (numerical & weighted share)

	No. of Industries + artisans	*%*	*Estimated 1979 fine gold cons. (tons)*	*%*	*Estimated 1980 fine gold cons. (tons)*	*%*
Arezzo	254	8	90	40	40	37
Milano	490	16	16	7	10	9
Valenza	1,095	37	30	13	20	19
Vicenza	520	18	67	30	30	28
Others	641	21	20	10	7	7
Total	3,000	100	223	100	107	100

more than the combined total of all other European countries. In that year, there were over 250 manufacturers in Arezzo, 520 in Vicenza and 1,000 in Valenza Po (Table 1).

The 1980 world crisis brought the Italian gold jewellery industry's steady expansion to an abrupt halt: due to the collapse of exports, mainly to Arab countries and the US, gold jewellery output dropped to 107 tons, decreasing by 53 percent over the previous year. However, last year there was a more positive trend and preliminary estimates for 1981 put total manufactured output at about 165 tons, 50 of which were absorbed by the domestic market and 115 exported. These results were achieved thanks to a relative stability of the gold price, increased promotional efforts by industry and trade, and the reopening of major markets for Italian exports, such as Arab countries, the United States and Latin America.

It is therefore encouraging to see that there are current signs of a recovery, and that manufacturers are looking forward confidently to 1982.

3. Factors for the Italian Leadership

But quite apart from any cold evaluation of consumption figures for fine gold, the importance that the Italian industry has assumed in the last decade is such that it has practically monopolized the market trends and is able to influence jewellery styling in a decisive manner with products made in Italy.

What then is the secret behind the success of the Italian gold jewellery industry? How can one explain the fact that in the last decade Italy managed to produce over 50 percent of all European gold jewellery, or almost 25 percent of the total world production (Tables 2 and 3).

Several factors have contributed to Italy's leadership in the gold jewellery sector:

— Tradition certainly plays an important part. Gold, and especially gold jewellery, has always had a place in the country's culture and civilization:

Table 2 - *Gold purchases for jewellery manufacturing*
(Italy versus the World; tons fine gold)

	1977	*%*	*1978*	*%*	*1979*	*%*	*1980*	*%*
Italy	209.0	21	235.0	23	227.0	31	107.0	30
Other Countries	793.0	79	772.0	77	510.0	69	247.0	70
Total W/W Production	1002.0	100	1007.0	100	737.0	100	354.0	100

Table 3 - *Gold purchases for jewellery manufacturing*
(Italy versus Europe; tons fine gold)

	1977	*%*	*1978*	*%*	*1979*	*%*	*1980*	*%*
Italy	210.0	51	235.0	53	227.0	55	107.0	50
Other European Countries	205.8	49	208.4	47	182.0	45	106.9	50
Total Europe	415.8	100	443.4	100	409.0	100	213.9	100

starting from ancient Roman times, through the Middle Ages and again during the Renaissance, when many famous painters and sculptors were trained as goldsmiths in their youth, Verrocchio, Ghiberti, Brunnelleschi, Antonio del Pollaiolo and, of course, Benvenuto Cellini our most renowned goldsmith.

The origin of the gold jewellery industry in the Vicenza area is firmly rooted in the goldsmiths' traditions of the Venetian Republic. While in Milan there are still street names which, to this day, commemorate the goldsmiths' guild, the connections betweeen Arezzo and Etruscan goldsmiths' traditions are obvious.

Tradition is therefore an important element in the Italian gold jewellery industry and provides a fertile background for its development.

— Another important element contributing to Italian supremacy in the jewellery field is the sector's advanced technology, mainly in chain-manufacturing machines. The Italians are unrivalled in this field and the efficiency of their chain-machines, a glorified version of a sewing machine as a brilliant journalist has defined them, is one of the main reasons for the leading position gained in the mass production of gold jewellery.

— We should also add that Italians have a particular talent for transforming raw materials into manufactured articles notable for their design, good taste, and high-quality finish. And Italian design in the gold jewellery field is second to none.

— Last but not least, another reason for the strength of the Italian gold jewellery industry is that the country's domestic market is one of the largest and most consistent in the world. In fact, Italians purchased gold jewellery amounting to 60 tons of fine gold in 1979 (Table 4).

In 1980 consumption dropped to 38 tons but it is estimated to have reached 50 tons in 1981. These figures place Italy at the top of the league in Europe and provide Italian manufacturers with a solid foundation on which to base their export drive.

Italy's domestic market is covered by over 12,000 gold jewellery outlets

Table 4 - *European domestic jewellery consumption* (tons fine gold)

	1977	*%*	*1978*	*%*	*1979*	*%*	*1980*	*%*
Italy	47.0	19	55.4	21	59.5	24	37.9	26
W. Germany	47.6	19	50.9	20	48.6	19	32.0	22
Spain	43.9	18	48.6	19	33.8	14	14.7	10
United Kingdom	27.0	11	27.1	11	30.5	12	14.3	10
France	30.6	12	29.7	12	27.3	11	15.8	11
Others	48.7	21	45.5	17	50.3	20	30.7	21
Total Europe	244.8	100	257.2	100	250.0	100	145.4	100

corresponding to about one retail outlet for every 3,600 adults, as compared with one for every 6-7,000 adults in other European countries.

A high percentage of the gold jewellery purchases in our country is linked to traditional gift occasions: Christening gold chains, First Holy Communion and Confirmation gold medals, engagement rings and wedding bands, all in 18 carat gold. Italians traditionally prefer high-carat gold jewellery, and sales of jewellery in alloys other than 18 carat are practically unknown.

A particularly interesting fact about the pattern of gold jewellery purchases in Italy is that they are inversely proportional to income: Area 4 Nielsen, which represents southern Italy, purchases more than 36 percent of the total amount of jewellery manufactured for the domestic market, and that is well above the percentage purchased in the richer industrial north.

4. Prospects for the Gold Jewellery Industry

With this in mind, what are the prospects for the gold jewellery industry in Italy in the 1980's?

The present leading position was achieved, until a few years ago, almost entirely by the individual efforts of manufacturers and/or distributors who succeeded in contacting buyers throughout the world. They did this the hard way, often using bold Garibaldian methods, and with little or no help from official or sectorial organizations.

At the same time it must be recognized that Italian leadership may not last forever. Soaring labour costs and industrial unrest have diminished our competitive edge and serious attempts are being made to start mass production of jewellery elsewhere, especially in cheap labour centres in the Far East.

It is now generally accepted that while the 1970-1979 decade favoured a production-oriented approach, in the 1980's the gold jewellery industry will have to place more emphasis on its marketing tools in order to seize new market opportunities and outlets in an ever-changing environment. Fluctuations in the gold price, a still uncertain economic situation worldwide and increasing competition from alternative luxury items will force jewellery manufactures to become more marketing-oriented.

Since the gold jewellery industry, both in Italy and abroad, is limited in size and extent and therefore unable to create adequate internal structures, we think that one of the few chances available for maintaining or even increasing our market share is by improving the efficiency of group and trade associations and by creating cooperative structures able to provide their members with services and assistance which they would not be able to obtain in their single capacity.

In our opinion, the pioneer era of the gold jewellery industry definitely ended with the 1980 crisis. Fuller and more precise market intelligence, reasearch for new and more advanced technology, and the increasing importance of finance together stress the need for more "managerial" and professional guidance. And the gold jewellery industry, so discreet and "hermetic" all over the world, needs new inputs, a new flow of experiences gained in other fields.

The Italian gold jewellery industry will be able to defend its leading position provided it is able to modify its structures and adapt to changing market conditions. If it does this while maintaining the creative talent of its operators, its advanced technological level and its natural vigour, it will certainly retain and keep, perhaps even improve, its market share in the gold jewellery sector worldwide.

CHAPTER 14

THE GOLDEN CONSTANT, OR THE GOLD STANDARD AND THE BEHAVIOUR OF COMMODITY PRICE LEVELS

ROY W. JASTRAM

1. The Relevance of Statistical Analysis

This meeting for which we are gathered is but one manifestation of the expanding worldwide interest in gold: its possible use as a means of stabilizing currencies and restoring stability to domestic and international price levels.

We have a great deal of evidence of this interest in the United States; not only among economists, important segments of government, significant sectors of the financial community, but among the general citizenry as well. During my own career as a professor in private and public universities I can remember no other time when there was such informed public discussion of monetary affairs as there is now, much of it centered around means of ending the inflation which, at one rate or another, has characterized the last fifty years in the United States.

I am here today as an analyst, not as an advocate of a single point of view. My historical researches have led me to see the stabilizing powers of gold within a monetary system. But I certainly do not take the view that a return of the gold discipline would cure all our economic ills.

As an analyst, I would like to show you some statistical conclusions I have reached based upon years of research leading to two books on the precious metals, *The Golden Constant* and *Silver: The Restless Metal*, John Wiley & Sons, New York.

Drawing upon my research on gold, particularly, I have placed before you two-charts, one on the price history of England and another for the United States.

The longest continuous history we have of a monetary system based on a consistent form of gold standard is in England from 1717 to 1914. It was a gold coin standard, based on a steady official price per ounce of gold at 3 pounds, 17 shillings, $10^1/_2$ pence, for all of those 200 years. The essential feature of convertibility continued until 1931.

The second longest record of a consistent form of gold standard is in the United States from 1834 until 1914. During that entire period the official price of gold was $20.67 per ounce. Because of their longevity I will concentrate on the tenure of the gold standard in England and in the United States.

2. Two Hundred Years of English Experience

Figure 1 shows three statistical series for England that are germane to the topic of this paper. They are identified on the chart as the price of gold (so labled), the wholesale commodity price index number (CP) and the purchasing power of gold at the wholesale price level (PPG). The last measures the exchange rate between an ounce of gold and commodities generally over the two centuries of the classical gold standard in England. All three series are indexed on the base 1930 = 100.0[1].

There are two striking features of the chart:

1) For 200 years commodity prices moved along a level plane, centered on the fixed price of gold.[2]

2) As a result, the purchasing power of gold showed no secular trend for two centuries.

An ounce of gold would buy in the commodity market of 1717 approximately the equivalent it would buy at wholesale in 1914. What is more, the cycles of commodity price in between were upward and downward around the stable price of gold in such a way that they averaged out very closely to zero over the whole span of time.

This is what a classical gold standard is supposed to do in theory. But that it did its work in practice over so long a period, and so closely in accord with monetary theory, is the truly remarkable aspect of my empirical findings. England went from a pastoral nation, largely dependent on agriculture, to an industrial economy dominating world commerce and finance. She ran the full scale from labour-intensive to capital-intensive economics. In the financial sphere she operated before the invention of credit currency, when only coins were used as a common denominator for exchange, and

---

[1] For a full explanation of how the original data were obtained and how the index numbers were calculated, see R. W. Jastram, *The Golden Constant,* John Wiley & Sons, New York, 1977, especially Chapters 1 and 3.

[2] The only time when the price of gold was not fixed was from 1797 to 1821 when specie redemption was suspended largely because of the Napoleonic Wars. As the chart shows, the temporary abrogation of the gold standard was accompanied by an unprecedented commodity price inflation.

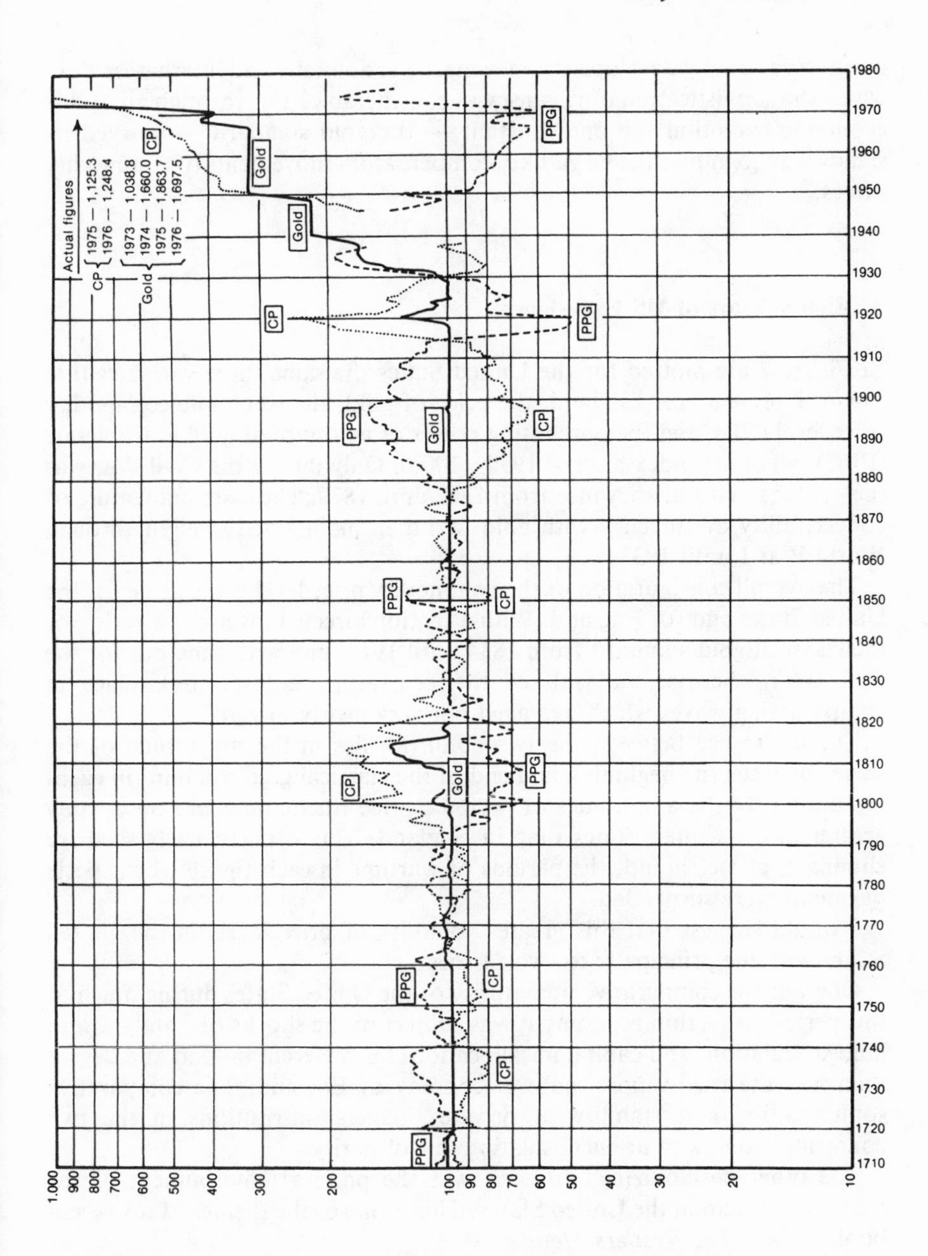

Fig. 1 The English experience: indexes of the price of gold, commodities, and purchasing power, 1560-1976: 1930 = 100.0

Source: Roy W. Jastram, *The Golden Constant*, John Wiley & Sons, Inc., 1977.

progressed to fully developed money markets, domestic and international, in which the sophistication of finance was considerable. Yet, through all of this economic evolution the one constant — the gold standard — worked its steady way to induce the wave-like, compensatory movements of commodity prices.

3. Eighty Years of US Experience

In Figure 2 are plotted for the United States the same three variables that Figure 1 presents for England: the price of gold, the wholesale commodity price level (CP), and the purchasing power of an ounce of gold at wholesale (PPG), all on an index base of 1930 = 100.0. Only during the Civil War was the gold standard in abeyance from 1862 until 1879. The essential feature of convertibility of currency with gold was then maintained straight through World War I until 1933.

The overall configuration of the commodity price level is the same for the United States and for England. With attention directed toward the period of the classical gold standard from 1834 until 1914, and with time out for the Civil War, the rise and fall of the general price level took place in compensating waves which averaged out very nearly to zero.

The difference between the two countries lies in the magnitude of the waves between the beginning and end of the classical gold standard in each. Up until 1914 the amplitudes of periodic price fluctuations are observably greater in the United States than in England. This is particularly so if we eliminate, as we should, the periods of wartime in each figure when specie payments were suspended.

I would suggest that this greater volatility of price levels in the United States was due principally to two factors.

One was the comparative immaturity of the United States during much of this period. As a thin economy it was subject to the shocks of immigration, land speculation, and capital misallocation, far different in kind and degree than was the vastly more mature economy of England. The comparative sophistication and stability of private financial institutions in the two countries is one way of encapsulating this disparity.

The other destabilizing influence was the political environment of the monetary system in the United States. This is one of the themes of my recent book, *Silver: The Restless Metal.*

The political passion surrounding the post-Civil War monetary debates alone can justify this characterization.

The inherent stability of a gold standard monetary system was constantly whipsawed by political maneuverings as a consequence. The ready tool of silver gave financial focus to the diverse political factions behind it, ranging

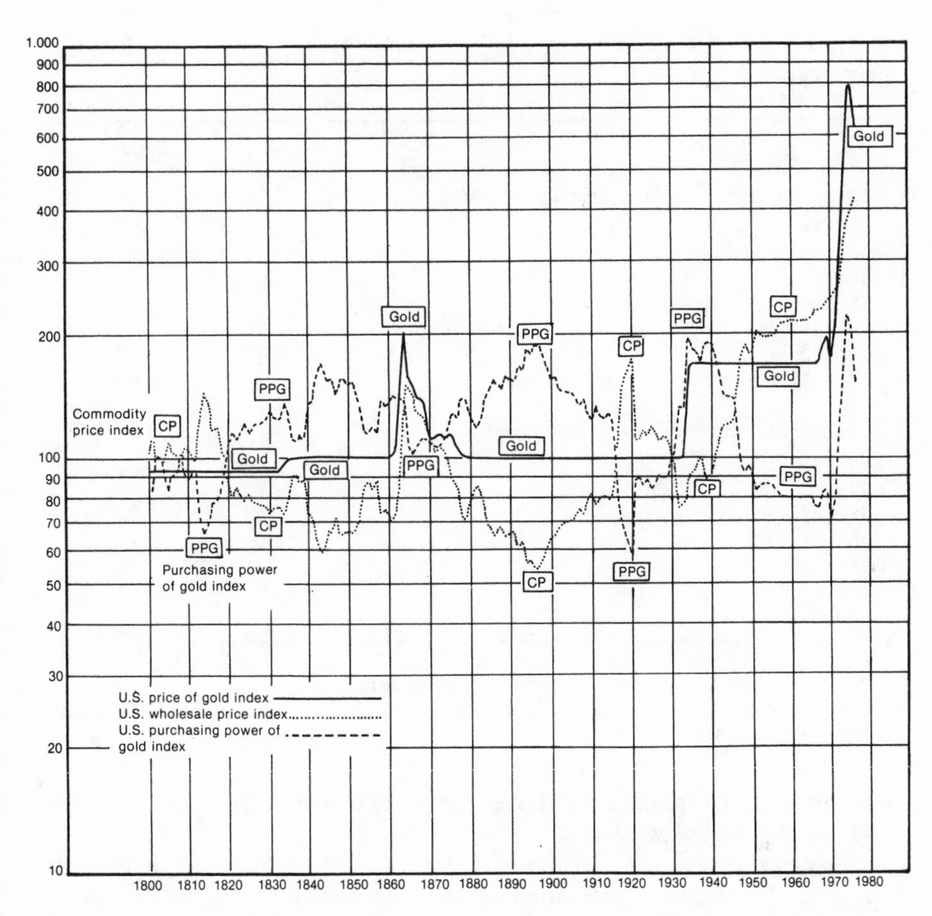

Fig. 2 The American experience: indexes of the price of gold, commodities, and purchasing power, 1800-1976: 1930 = 100.0

Source: Roy W. Jastram, *The Golden Constant*, John Wiley & Sons, Inc., 1977.

from the agrarians of the central states to the mining interest of the West.

4. The Stabilizing Evidence of the Classical Gold Standard

Throughout this turbulence the classical gold standard worked its long run steadying influence in the United States just as it did in England. Only the fluctuations around a level plane were greater in the United States for reasons that were in large part political rather than economic.

Figures 1 and 2 display the phenomenon of the gold discipline in graphic form. Table 1, which follows, is a convincing demonstration in arithmetic

Table 1 - *Change in the wholesale price index* (in percent)

Years	*Inflation*	*Deflation*
United States (net sum over time equals —3%):		
1808-14	+53.1	
1814-30		—72.2
1843-61*	+11.2	
* 1879-97		—17.5
1897-1914**	+24.9	
1914-20	+100.0	
1920-33		—102.5
England (net sum over time equals +0.6%):		
1752-76	+22.1	
1792-97*	+19.5	
* 1820-43		—38.5
1843-57	+22.1	
1857-96		—45.3
1896-1920	+195.9	
1920-31		—175.2

* Years between which gold convertibility was suspended.
** This inflationary period is separated at 1914 because that is the year the Federal Reserve System came into operation.

that the gold discipline worked over the years of *convertibility* in England and the United States.[3]

The years represent the periods of major inflation and deflation in the two countries. The extent of inflation and deflation is measured by the change in percentage points of the wholesale price index numbers. For example, a decline in the wholesale price index from 150.0 to 95.0 would be registered as a change of -55 percentage points. Years when the gold standard was suspended in each country are omitted from the table as indicated by the asterisks.

We see that over a combined record of about 300 years these gold standard countries had inflations and deflations that averaged out sensibly to zero. More strikingly, the index of wholesale prices in England was at the same figure in 1930 — the last full year of convertibility — as it was in 1717 when the gold standard began.

[3] The years of "convertibility" extend to 1931 and 1933 for England and the United States, respectively. The years of the "classical" gold standard terminate with 1914 for both: in England because the freedom of gold export and import was terminated then; in the United States because of the inauguration of the Federal Reserve System at that time.

Major swings in the commodity price level certainly did take place under the gold standard, but then turned out be compensatory in nature and magnitude.

In both England and the United States they averaged out to zero, and the wholesale commodity price index wound up in 1930 just where it began in 1717 and 1800, respectively, in the two countries.

A final way of capturing statistically the stable relations involved is to present a table showing the purchasing power of an ounce of gold at wholesale commodity price levels over a very long span of time. To do this we can draw upon the purchasing power of gold index (PPG) pictured graphically in Figure 1 for England.

The following Table 2 starts with the inauguration of the gold standard in 1717 and proceeds by decades (approximately) until England went off convertibility in 1931. The guiding criterion is the convertibility of the pound into gold; wartime suspensions of specie payments are omitted, such as between 1790 and 1822. The stability displayed is remarkable.

Table 2 - *Index of the purchasing power of gold: England, 1717-1930* (1930 = 100.0)

Year	*Index*	*Year*	*Index*
1717	103.2	1840	90.2
1723	98.5	1850	125.7
1730	111.4	1860	97.7
1740	99.4	1870	100.8
1750	112.7	1880	110.0
1760	110.8	1890	134.5
1770	110.7	1900	129.2
1780	112.7	1910	124.1
* 1790	103.1	* 1914	113.9
* 1822	104.6	1930	100.0
1830	97.7		

5. Lessons from the Past and the Present Situation

There is considerable interest at the present time in moving to some form of gold standard to recapture a degree of stability in the price level that has not been experienced for some 50 years. Opponents assert that the economic and social context has changed greatly since the time of the classical gold standard; that whatever the lessons learned then, they are not applicable now. But the economic and social circumstances in England changed hugely from the early 18th century until the end of the 19th into the 20th. And in the United States the decades leading up to the First World War were vastly different from the times when that nation's economy was in early consolidation.

Surely we should not reject out-of-hand the power of a gold standard to establish once again a long run stability of commodity price levels. Especially when we now have our present advantages in the degree of sophistication in finance and the benefits of high-technology in near-instant communication.

Let me close with some startling statistics.

From the time the United States went off the gold standard in 1933 the wholesale price level has gone up by 760 percent. Since England abrogated the gold standard in 1931 her price index number has risen by over 2000 percent.

Before that the two countries had a combined history of nearly 350 years of long run price stability. The price level was the same in the United States in 1930 as it had been in 1800. In England the price index stood at 100.0 in 1717 (the first year of her gold standard) and it was at that figure again in 1930.

Some final thoughts of mine were generated by this long historical record:

1) There must be discipline over the money supply. Nearly everyone agrees with this in the abstract. Disagreement arises over the question of *at what levels* and *how* to exercise the discipline.
2) Attempts at monetary discipline when managed by men have not worked. The *only* exceptions were draconian measures ending brief periods of crisis.
3) Therefore I believe there must be management by law, not by men.
4) Those monetary laws that worked best throughout history have been based upon the discipline of the precious metals. Notice that I am not saying that whenever the system was based on precious metals it was stable; I am saying when in history we find long run stability of prices we find precious metals standing behind it.
5) The precious metal that has had the most successful experience in stabilizing price levels is gold.

CHAPTER 15

INFLATIONARY EXPECTATIONS AND THE PRICE OF GOLD

ROBERT Z. ALIBER*

1. The Problem of the Appropriate Monetary Price for Gold

A major concern in any new role for gold as a domestic money or as an international money is the appropriate monetary price for gold — the price of a standard unit of gold, an ounce or gram, in terms of the standard monetary unit, such as the dollar or pound. One purpose of any new monetary arrangement based on gold is to achieve greater stability in the price level by increased reliance on an automatic mechanism for determining the rate of growth of money and reduced reliance on discretionary arrangements. A second purpose is to reduce variability in exchange rates by strengthening international reserve arrangements; central banks would have greater assurance about the foreign exchange equivalent of their gold holdings.

In the nineteenth century and earlier, gold standard arrangements developed to protect holders of money from the failure or bankruptcy of the money producing institutions. Subsequently, a rationale developed that the gold standard would assure stability of the world commodity price level because of the "niggardliness of nature." Nevertheless, long run price level stability appeared to result from the homeostatic properties of the monetary mechanism. If the chance of gold discoveries led to an unanticipated increase in the money supply, the subsequent increase in the world commodity price level would reduce gold production and encourage the non-monetary commodity demand for gold. The rate of growth of the money supply would decline, and, with an unchanged rate of growth of the demand for money, the commodity price level would decline. The argument that there was an equilibrium relationship between the monetary price of gold and the

---

* Cole Kendall has been helpful with the development of the data in this paper.

commodity price level was based on the responsiveness of gold output and of the non-monetary demand for gold to changes in the price level. Despite this argument, price level stability did not become an important objective of policy until the 1920's. Somehow the argument about the price level stabilizing effects of a gold standard arrangement became transformed into the belief that the gold provided a hedge against inflationary increases in the consumer price level.

The historical argument for a return to a gold standard is based on stability of world price level during the nineteenth century. There have been other periods when the world price level showed far less stability than this example suggests, even though gold was the dominant money: for example, in the sixteenth century, the price level rose by a factor of five as a result of Spanish conquest of the New World — or at the rate of 1.9 percent a year. A primary question is whether the long run price stability in the nineteenth century was inherent in the gold arrangements or an historical fluke.

Casual observation suggests that the market price of gold since the late 1970's has been substantially higher than the estimate of the long run equilibrium price inferred from increases in the world commodity price level. The market price of gold has increased by a factor of ten, while the world price level has gone up two to three times. Either the recent market prices of gold contain a substantial "inflation" premium, loosely comparable to the Fisher premium built into interest rates, or instead there has been a change in the "equilibrium relationship" between the monetary price of gold and the world commodity price level. If this relationship has changed — or is continuing to change — then the proposition that a return to gold will be associated with price level stability is questionable.

The next section of this paper discusses the returns to the ownership of gold and the long run relationship between the monetary price of gold and the world commodity price level. Then the subsequent section seeks to explain recent changes in the dollar price of gold in terms of inflationary expectations since the dollar-gold link was broken in 1971. The concluding section seeks to reconcile recent levels of the gold price with the long run equilibrium price for gold.

2. The Long Run Equilibrium Price of Gold

The argument for a gold standard is that world commodity price level stability can be more nearly assured by adherence to some monetary rule based on gold flows than by discretionary monetary management. Thus, as long as the monetary price of gold remains unchanged, the world commodity price level should not change significantly in the long run. The implicit assumption is that increases in the demand for money arising from increases in the level of real income would be accommodated through increases in world gold output, and perhaps by reductions in the non-monetary demand. Increases in the commodity price levels attributable to random shocks, perhaps to new gold discoveries, would be followed by decreases in the world commodity price level from declines in the money supply following a decline in gold output because of the increase in gold mining costs. The tangential argument, that gold is a hedge against inflation, implies that the price of gold increases whenever there is a significant increase in the commodity price level; this argument also assumes that there is a long run equilibrium relationship between the market price of gold and the commodity price level.

Change in the price of gold can be expressed as a rate of return, both as a monetary rate of return in nominal terms and as a real or price level adjusted rate of return. The monetary return is the return from owning gold in terms of the other monies, especially national monies like the US dollar, the British pound, and the Swiss franc; this return can be measured as the average annual increase in its price in terms of one of these national monies. The real return is the nominal return deflated by either changes in world commodity price level or by a national price level.

The view that gold standard leads to stability of the world commodity price level implies that the monetary rate of return from owning gold is zero. Nevertheless, private parties hold gold because it offers a non-monetary convenience return; indeed, the value of this return is at least as high as the interest rate that might have been earned on the riskless asset — or perhaps on an asset in the same risk class as gold. Investors appear to attach a higher convenience return to gold than to other commodities. The larger this non-monetary return, the smaller the monetary return required to induce investors to hold gold.

If the commodity price level is more or less unchanged in the long run, there should be no significant difference between the monetary return and the real return attached to gold ownership. If the commodity price level decreases, the real return may be negative even while the monetary return is unchanged. If gold is an inflationary hedge, the monetary price may increase; however, the real return would not differ significantly from zero, especially if gold is a perfect inflationary hedge. There appears to be very little in the "theory of the gold standard" or even in any popular interpretation of the gold standard that would justify a positive real return.

For several centuries the monetary return from holding gold was zero, since Great Britain mantained the same gold parity until 1931, with several wartime exceptions. For a period approaching 200 years, there was no monetary return, and the real return was less than one-tenth of one percent a year. The US gold parity was established (or inherited) by the Act of Coinage in 1791, and this parity remained effective until 1933, except for a modest change in the gold content of the dollar in the 1830's. The average annual absolute change in the US price level over this period was only several tenths of one percent. Even though there may have been a disequilibrium relationship between the monetary price of gold and the commodity price level at the end of the period (or even at the beginning), adjustments for this disequilibrium would not have a major impact on the average annual return because of the extended length of the period. Thus, even if there had been a disequilibrium of twenty percent or thirty percent at either the beginning or the end of the period, the impact on the average annual change in the real return would be in the order of one-tenth to two-tenths of one percent a year. Hence, for several centuries the monetary return from holding gold was zero, and the real return was slightly negative — but only by several tenths of one percent.

The monetary return from owning gold increased significantly in the interwar period, for the price of gold in terms of the dollar and most other currencies was increased by about seventy-five percent in the 1930's. One interpretation of the changes in mint parities in the 1930's was that they were a belated response to a gold shortage that resulted from the increase in the world commodity price level during and after World War I — a concern that led to the conferences in Brussels and Genoa in the early 1920's, and to proposals to economize on the use of gold payments by private parties and in central bank reserves. Thus the US commodity price level in 1929 was 40 percent higher than the price level at the beginning of World War I; according to gold standard theory, the private demand for gold would be higher and gold production would decline leading to a gold shortage. An alternative interpretation of the 1930's increase in parities, is based on the domino-like succession of currency devaluations, when individual countries devalued to avoid a worsening in their international competitive position from the prior devaluations of Great Britain, which was the first of the major countries to go off gold (several Latin American countries had devalued as much as two years earlier), in part to rectify the inappropriate alignment of parities established in the mid-1920's. Thus, in 1925 the British should have returned to gold at the parity which might have resulted in a dollar-sterling rate of $4.50 rather than $4.86. When the French again pegged to gold at the end of 1926, the franc was undervalued. After the British stopped pegging to gold, sterling depreciated and speculative pressure was directed at the dollar and other currencies still pegged to gold. The increase in the price of gold to $35 reflected domestic monetary considerations, especially a desire to reflate the US price level. The evidence against the view that there was a gold shortage in the 1920's is that gold output was only modestly lower than in the years before World War I, nor did the private commodity demand for gold seem to be significantly larger. If private parties contemplated a gold shortage, they might have increased their demand for gold in anticipation of an eventual increase in its price, as in the 1960's and the 1970's.

If the theory of the gold standard arrangements is valid, then the equilibrium price of gold in the 1980's should be related to the change in the world commodity price levels since the 1960's or 1950's or some earlier period, when an equilibrium relationship prevailed between the monetary price of gold and the commodity price level. The financial implication of this theory is that the nominal rate of return from owning gold should not differ significantly from zero whenever the commodity price level is more or less unchanged. If, however, there is a significant increase in the commodity

Table 1 - *Predicted gold parities**

enchmark ate		*Predicted 1982 Parities*	*Rates of Return*		*Range for 1982 Parities*		
			Nominal	*Real*	*± 1%*	*± 2%*	*± 3%*
		(1)	(2)	(3)	(4)	(5)	(6)
975	US $	298	13.8	4.3	279-317	262-338	249-360
	SF	528	8.0	4.5	494-565	461-604	430-645
		($203)			($190-$217)	($177-$232)	($165-$248)
965	US $	110	15.4	7.9	94-129	80-151	68-176
	SF	325	9.6	4.8	276-382	234-449	198-526
		($76)			($64-$89)	($54-$104)	($46-$122)
955	US $	$130	9.4	4.2	100-168	77-216	59-278
	SF	405	6.0	2.2	311-524	239-678	182-874
		($94)			($72-$122)	($56-$158)	($43-$203)
945	US $	193	6.8	2.0	136-275	95-389	66-550
	SF	472	4.3	1.1	329-674	229-960	158-1363
		($110)			($77-$157)	($53-$223)	($37-$317)
935	US $	253	5.3	1.0	161-397	102-619	64-961
	SF	606	4.1	0.4	385-952	243-1488	153-2316
		($195)			($124-$307)	($78-$480)	($49-$747)
925	US $	117	5.3	2.2	67-203	38-350	22-601
	SF	455	3.3	0.8	260-791	148-1368	84-2353
		($88)			($50-$152)	($28-$263)	($16-$453)
915	US $	202	4.5	1.0	106-386	55-730	28-1374
905	US $	267	3.9	0.5	126-560	59-890	28-2416

$$\text{Predicted Parity} = \text{Parity at Benchmark Dates} \times \frac{\text{Price Level 1981}}{\text{Price Level Benchmark Date}}$$

* Assuming 1982 price of gold = $400 = SF 720
Swiss franc data converted to US dollars at current exchange rates.

Source: IMF, *International Financial Statistics*.

price level, the monetary price may increase, and the monetary return may match the increase in the commodity price level. Even then, however, the real return should not differ significantly from zero.

The appropriate 1982 "parity" for gold in terms of the dollar and the Swiss franc can be inferred from the increases in the commodity price levels from alternative benchmark dates (these parities are selected so that the real rate of return is not significantly different from zero). Thus column (1) in Table 1 shows the "equilibrium" 1982 parities for gold both in terms of the dollar and the Swiss franc on the assumption that 1982 parities should be directly proportional to the increase in the US and Swiss price levels since the various benchmark dates. The monetary rate of return from holding gold from each benchmark date to the end of 1981 is shown in column (2), while column (3) shows the real rate of return, which is the monetary rate of return deflated by the average annual increase in each country's price level. Columns (4) and (5) show the range for the 1982 equilibrium price of gold in terms of both the dollar and the Swiss franc, on the assumption there might be a modest average annual deviation between the increase in the monetary price of gold since the benchmark date and the increase in the commodity price levels. In effect, these two rows indicate the range for the 1982 parities on the assumption that the real rates of return might be modestly positive or negative.

By almost any measure, the recent market prices of gold are high relative to the predicted parities from any benchmark date. For example, if 1935 is the benchmark date, the appropriate 1982 US dollar price is $253; if 1945 is the benchmark date, the appropriate 1982 price is $193. Moreover, the market price for gold in terms of the Swiss franc is also high relative to the predicted price, although, because of the real appreciation of the Swiss franc relative to the US dollar, the gap between the market price of gold and the predicted "parity" is smaller when gold is priced in terms of the Swiss franc. If the view that there is a one-to-one relationship between changes in commodity price levels and the monetary price of gold is relaxed modestly to permit cumulative deviations between the world price level and the monetary price of gold, of one and two percent a year, and the benchmark date is sufficiently distant, then recent market prices of gold fall within the acceptable range. However, the view that such small cumulative deviations are consistent with the theory of the gold standard as a price-stabilizing mechanism is questionable. Over a century a two percent average annual cumulative deviation would lead to an increase in the price level by a factor of seven to eight.

The nominal and real returns in terms of the US dollar and the Swiss franc are shown in Tables 2A and 2B. For the longest possible period (1905-1982 for the US dollar and 1925-1982 for the Swiss franc), the nominal returns are 3.9 percent a year and 3.4 percent a year, and the real returns are 0.5 per cent and 0.8 percent a year. These real returns might not seem inconsistent with

Table 2A - *Returns from holding gold, US dollars*

	Base Years							
	1905	*1915*	*1925*	*1935*	*1945*	*1955*	*1965*	*1975*
Terminal Year								
				Nominal Returns (percent/year)				
1915	0.0							
1925	0.0	0.0						
1935	1.8	2.7	5.4					
1945	1.3	1.8	2.7	0.0				
1955	1.1	1.3	1.8	0.0	0.0			
1965	0.9	1.1	1.3	0.0	0.0	0.0		
1975	3.0	3.5	4.2	3.9	5.2	7.9	16.5	
1982	3.9	4.5	5.3	5.3	6.8	9.4	15.4	13.6
				Real Returns (percent/year)				
1915	—2.7							
1925	—4.0	—5.3						
1935	—0.2	1.1	8.0					
1940	—0.8	—0.2	2.5	—2.7				
1955	—1.4	—1.1	0.3	—3.3	—3.9			
1965	—1.5	—1.2	—0.2	—2.7	—2.8	—1.6		
1975	0.2	0.6	1.9	0.4	1.4	4.2	10.4	
1982	0.5	1.0	2.2	1.0	2.0	4.7	7.9	4.3

Source: IMF, *International Financial Statistics*

Table 2B - *Returns from holding gold, Swiss francs*

	Base Years					
	1925	*1935*	*1945*	*1955*	*1965*	*1975*
Terminal Year						
				Nominal Returns (percent/year)		
1935	0.0					
1945	1.7	3.4				
1955	1.1	1.7	0.0			
1965	0.8	1.1	0.0	0.0		
1975	2.8	3.5	3.5	5.3	10.8	
1982	3.4	4.1	4.3	6.0	9.6	8.0
				Real Returns (percent/year)		
1935	2.9					
1945	6.2	—2.5				
1955	—0.4	—2.0	—1.5			
1965	—0.8	—2.0	—1.8	—2.1		
1975	0.3	—0.3	0.4	1.3	5.0	
1982	0.8	0.4	1.1	2.2	4.8	4.5

the theory of the gold standard. Thus the nominal returns and the real returns in the 1965-75 decade especially might be considered exceptional, and perhaps attributable to a "catching-up" phenomenon. Nevertheless while real returns attached to ownership of gold in the twentieth century have resulted in real returns of less than one percent a year, these real returns in the twentieth century were several times larger than the real returns in previous centuries.

3. The Market Price of Gold in the Short Run

The sharp increase in the price of gold in the 1970's, both in terms of the dollar and in terms of the Swiss franc, was largely unanticipated. This price increase has no good historical precedent, for in the past, the stability of the relationship between the monetary price of gold and the commodity price level reflected the variations in the commodity price level rather than increases or decreases in the gold price. In the 1965-75 decade, the nominal return from holding gold was 16.5 percent in terms of the dollar and 10.8 percent in terms of the Swiss franc; over the more extended 1965-82 periods, the average annual returns were 15.4 and 9.6 percent a year, respectively. The real returns were 7.9 percent for gold in terms of the US dollar, and 4.8 percent in terms of the Swiss franc. While interest rates in US Treasury bills were higher during this period than in the previous decades, the real returns associated with owning gold were substantially higher than the real returns from holding Treasury bills. Hence, the traditional relationship that equilibrium involved a higher interest rate on Treasury bills than on gold, because gold had a non-monetary convenience yield, had been reversed. Whether the higher real return on gold than in the 1970's reflects that gold has become a riskier asset now that its price is no longer pegged remains conjectural. What remains to be determined is whether the return from holding gold is appropriate in terms of the risk. During the 1970's the trend in gold price movements was important; the price had doubled between three and four times. A number of efforts have been made to explain the changes in the price of gold by equations like (1) where the price of gold is expressed as a function of time. Thus,

$$\log P_{gold} = \underset{(-0.05)}{-0.31} + \underset{(7.76)}{.059}\,\text{Trend};\; \bar{R}^2 = .622,\ \text{D-W} = 1.62,\quad \hat{\rho} = \underset{(13.5)}{0.87} \qquad (1)$$

for the period 1970 I to 1981 II. The same equation is applied to the data for the previous decades; however, the results are different. The coefficient on the time variable would have a value of zero. Equations like (1) have no economic content.

At the level of stylized facts, the peaks in the price of gold at the end of

1974 and early in 1980 more or less coincide with peaks in the rate of increase in the commodity price level. Increases in the money interest rates might be associated with decreases in gold price on the rationale that the opportunity cost of holding gold has increased. Changes in the anticipated price level and in the money interest rate combined to reflect the anticipated real interest rate; the market price of gold might be expected to increase when the anticipated real interest rate declines.

Thus

$$\log P_{gold} = a + b \frac{r}{\frac{P^*}{P}}$$

where r is the nominal interest rate

and $\frac{P^*}{P}$ is the anticipated rate of change of the price level.

Thus,

$$\log P_{gold} = \underset{(0.10)}{0.06} + \underset{(1.31)}{2.36} \frac{r\$}{\dot{p}/p} + \underset{(7.22)}{.056}\ \text{Trend} \qquad (2)$$

$$\bar{R}^2 = .627 \qquad \text{D-W} = 1.54 \qquad \hat{\rho} = \underset{(12.90)}{0.87}$$

Equation (2) has modestly more explanatory power than equation (1). While the coefficient on the real interest rate is substantially higher than the "theory" predicts, the coefficient does not appear to be significant.

Monthly data were used to estimate the response of changes in the gold price to changes in the real interest rate.

$$\Delta \log P_{gold} = \underset{(3.14)}{0.023} - \underset{(1.71)}{0.311}\ \Delta \frac{r\$}{\dot{p}/p}; \qquad \bar{R}^2 = .016 \qquad \text{D-W} = 1.766 \qquad (3)$$

The coefficient is lower than the theory predicts, but the coefficient is significant. The low $\bar{R}^2$ indicates that changes in the real interest rate explain only a very small part of changes in the market price of gold.

Observation of the movements in the gold price suggests that when the rate of change in the price level begins to increase the increase in price of gold initially is modest. As rate of change in the price of gold continues to increase, however, the price of gold begins to increase at a more rapid rate. It seems not unlikely that part of the increase in the price of gold may be due to a "bubble"; investors buy gold because they anticipate the price will increase because other investors will buy gold in the anticipation that the price will increase, etc. For brief limited periods, the increase in the gold price may be detached from increases in the commodity price level. Efforts to obtain a coefficient for the real interest rate nearer to one by truncating the period of observation were not successful.

In early 1982, the increase in the commodity price level was about seven percent a year, at an interest rate of thirteen percent, the real interest rate is about six percent. The implication is that the cost of holding gold is unusually high, at least by historical standards, and so the market price of gold might continue to fall.

4. Reconciling the Long Run and Equilibrium Prices of Gold

Recent market prices of gold reflect two terms — one is the long run equilibrium price associated with increases in the commodity price level, and the other is the anticipated change in the price level. If the price of gold is again pegged, the price must be sufficiently high so that private demand for gold for all possible uses is smaller than the new supply, unless the authorities wish to sell gold. The authorities might take one of several approaches to pegging the market price of gold.

— They might peg the price at or near the market price on July 4, 1982, or any other arbitrary date, and indicate that they are determined to maintain the parity.

— They might peg the price of gold on July 4, 1982 or any other arbitrary date, and indicate that they will change the peg if they are obliged to buy too large a volume of gold to maintain the peg.

— The authorities might wait until they believe the market price is at or near the long run equilibrium price before they peg. Estimates of the long run equilibrium price differ, and vary with the anticipated rate of increase in the commodity price level.

The long run equilibrium price is not independent of whether the authorities peg the price. The stronger the commitment of the authorities to maintaining the parity, the lower the long run equilibrium price of gold. The story is that the private demand for gold is smaller, the stronger the commitment of the authorities to maintaining the parity, since the anticipated monetary return is lower.

If at the time of pegging the long run equilibrium price is significantly below the market price, the authorities incur the risk that they may be obliged to purchase significant amounts of gold, both from existing holdings as well as from new production. The consequence is that they may be obliged to expand the monetary base and the money supply more rapidly than they prefer, or alternatively, that open market sales of securities would be necessary to offset open market purchases of gold. If instead they wait until the market price of gold falls to the level they believe most nearly approximate to the long run equilibrium price, they may fail to catch the turning point — that the market price may not fall to the authorities' estimates of the long run equilibrium price.

If the authorities adopt the "crawling parity" approach, and change the parity to limit their gold transactions, the impact on the private demand is likely to be small. The shortcoming of this approach is that the new policy is not likely to convince many that the authorities will alter their monetary policy to maintain the parity. Inevitably investors will believe that if the authorities adjust to the excess supply of gold by reducing their buying price, they will adjust to the excess demand by rising the selling price. Any policy that projects possible changes in the gold price to limit the transactions of the authorities is not likely to have a significant impact on the private demand for gold.

CHAPTER 16

ON THE INSTABILITY OF RAW MATERIALS PRICES AND THE PROBLEM OF GOLD

PAOLO SYLOS-LABINI

1. Economic Theory and Price Determination

I suggest that three groups of economists may be distinguished: 1) those who assume flexible prices in relation to demand, e.g., the Walrasians and the monetarists; 2) those who assume rigid prices, again, in relation to demand and up to the level of full employment, e.g., the Keynesians; and 3) those who feel that it is essential to distinguish between different categories of prices and wages, a group to which I belong.

More precisely, two dichotomies are to be considered, one in the commodity markets, the other in the labour markets. In the commodity market we must distinguish between the raw materials markets and the markets of manufactures. As I suggested some years ago, the former dichotomy must be related, not only to the market forms, but also to the technical conditions of production and to the type of goods. Arthur Okun has recently put forward again this dichotomy, relating it to characteristics of the goods (homogeneous or differentiated) rather than to the market forms, with producers who are "price takers" or "price makers" — a dichotomy which, however, presupposes different types of markets. This dichotomy corresponds to the one put forward in 1943 by Michael Kalecki (prices determined by demand and prices determined by costs), and, more recently, by Sir John Hicks (who distinguishes between "flexprices" and "fixprices"), by Lord Kaldor and by Joan Robinson.

As for the labour markets, we must distinguish between so-called manual labour (production workers, receiving wages) and the so-called intellectual labour (white collar workers, receiving salaries). In our times demand for labour has a limited influence on changes of wages and almost no influence on changes of salaries, at least in the short run; both wages and salaries are much influenced by the cost of living.

2. Price Determination: Raw Materials and Manufactured Products

As David Ricardo correctly observed in Chapter XXX of his *Principles,* prices depend on demand and supply under competitive conditions in the short run and, in monopoly, both in the short and in the long run, with the proviso — we may add — that in monopoly supply is administered in order to maximize profits in the long run, whereas this is not possible in competition. In the world markets, raw materials are supplied either under competitive conditions or (less frequently and only in the case of certain mineral products) under quasi monopolistic conditions.

The situation is different in manufacturing. Here, as a rule, we find oligopolistic conditions. In such conditions in the short run price does not depend on demand variations, but on variations of costs or, more precisely, of direct costs. This is basically due to three facts: 1) normally in manufacturing industry there is unused capacity, so that an increase in demand implies an increase in output rather than in prices, just as Keynes assumed; 2) direct costs do not vary with output, at least in the relevant range, as shown by an impressive list of empirical enquiries, usually neglected by economists; 3) the oligopolistic interdependence implies that each firm is willing to change its price only when conditions of the market change *for all;* in the short run changes in direct costs tend, in fact, to be

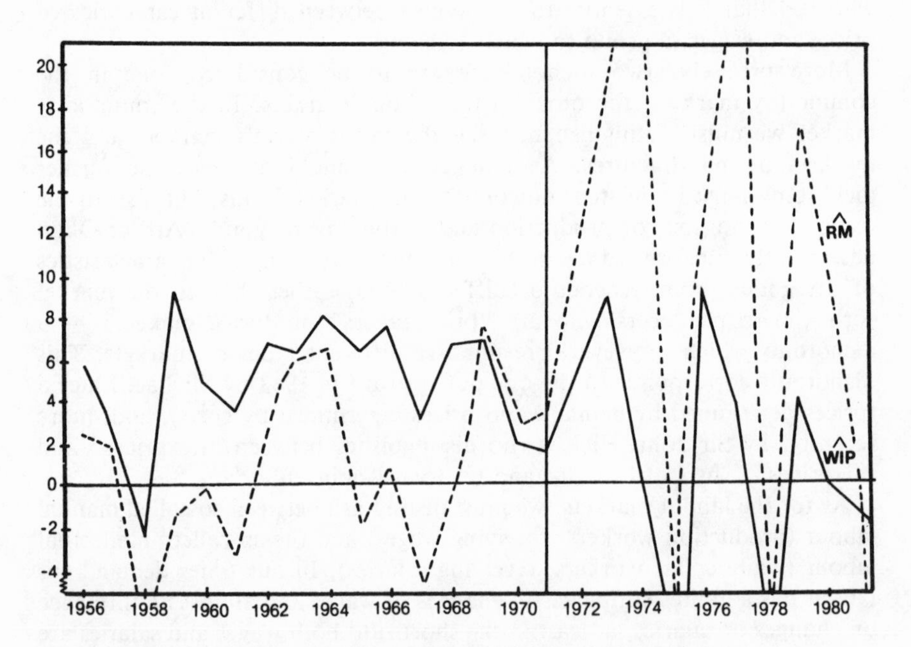

Fig. 1 Industrial production and prices of raw materials (rates of variation).

general. Demand variations can affect *directly* prices only in very special circumstances (like, for instance, in a situation of an international boom), or indirectly (and partially) via cost variations, or, again, indirectly, by affecting the balance of payments and the exchange rate.

In the past century competitive conditions were the rule in all markets, including industrial ones. In this century the process of concentration has asserted itself in several important industries and the commercial revolution, mainly through the development of means of transportation, of the mass media and advertising, has strongly increased the differentiation of products, transforming to a good extent price competition into quality competition. As a result, in manufacturing the rule is no longer competition in the proper sense, but oligopoly, in its three forms: concentrated, differentiated, and mixed. As a further consequence, it is wrong to expect prices to be flexible in relation to demand in the case of manufactures, whereas it is quite correct to expect prices to be flexible in the case of raw materials. In fact, this flexibility appears clearly in Figure 1 which refers to the period 1956-1980 and indicates the rates of variation of raw materials

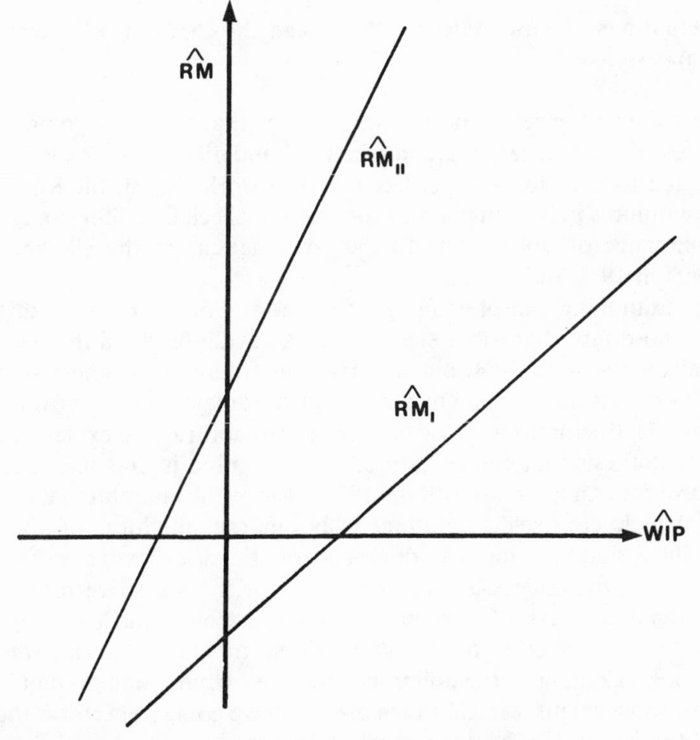

Fig. 2 Relation between rates of variation of industrial production and prices of raw materials: 1958-1971; 1972-1980.

prices ($\widehat{RM}$) and of the world industrial production ($\widehat{WIP}$) expressing the demand for raw materials.

From this graph another fact also emerges clearly, i.e. a sharp contrast between the behaviour of raw materials prices before and after 1971. More precisely, these prices appear to be flexible in relation to demand in both periods; but in the first period the rates of variation of raw materials prices are as a rule smaller than those of world industrial production, whereas after 1971 the extent of price fluctuations is 4 to 5 times greater; moreover, unlike what happened previously, price fluctuations are, for the most part, positive. Figure 2 illustrates the contrast between the two periods; the x-axis indicates the rates of variation of the world industrial production and the y-axis those of raw materials prices. Both the constant and the slope of the two regression lines are significantly different:

$$\text{I) (1958-1971)} \quad \widehat{RM} = -5.1 + 0.9\ \widehat{WIP}$$
$$\text{II) (1972-1980)} \quad \widehat{RM} = +9.1 + 2.4\ \widehat{WIP}$$

3. Fluctuations of Raw Materials Prices and the Crisis of the International Monetary System

A view quite widespread among economists is that the extraordinary rise in the prices of oil and raw materials in 1973 and 1974 is to be attributed to special and irrepeatable events, like the 1973 world boom, the Kippur war, the crop failures in certain parts of the world (especially USSR) and, finally, the abundance of dollars due to the large deficits in the US balance of payments in 1971 and 1972.

After examining the previous graph, one is bound to reject this view. There is no doubt that those special events have intensified the rise in raw materials prices in 1973-74: but that rise was followed by others — in 1977 and 1979 — less impressive but no less extraordinary when confronted with the pre-1971 fluctuations. How can this sharp contrast be explained?

The hypothesis that comes naturally to our mind is that such a contrast must have something to do with the 1971 crisis of the international monetary system which consisted, simultaneously, in the abolition of the dollar convertibily and in the abandonment of fixed exchange rates. More precisely, my hypothesis is that the ever-present speculative factor in the international markets of raw materials has become much stronger as a consequence of the crisis of the international monetary system, which has meant the weakening of the dollar as a reserve currency and accumulator of value; to some extent, certain raw materials have come to perform the latter role. Speculation has become much stronger not only in raw materials markets, but also in those of foreign exchanges after the adoption of flexible exchange rates. It is to be noted that the flexibility of exchange rates works asymmetrically, i.e. it works more easily in an upward than in a downward direction. This applies even more to speculation in the markets of raw materials. Speculation in foreign exchanges and speculation in raw materials markets often combine; it is possible to observe speculation in these markets even when the raw materials prices are expected to be stable if the dollar is rising in terms of the other currencies, since most of the transactions in raw materials markets are carried out in dollars. All this has contributed to determine a worldwide process of inflation.

4. The Role of Gold

In the new conditions, then, all prices have shown a systematic tendency to rise. In particular, the rise of raw materials prices has, via costs, pushed up the prices of manufactured goods and, in this way, the cost of living. The rise in the cost of living has pushed up wages, thus reinforcing, via costs, the increase in industrial prices. Such a process has taken place in all industrial countries, but the speed of the increase in prices and wages has been different in the different countries, depending on the structure of production, on public spending, on the strength of trade unions and on the type of institutions. In certain periods, for countries in deficit Triffin's vicious circle has occurred (rising exchange rates, rising prices and so on). In short, the worldwide inflation of the last ten years is to be attributed, to a large extent, directly or indirectly, to the crisis of the international monetary system; but the inflation differentials depend on the particular conditions of the different countries.

If this hypothesis is well founded, the way out of the strong inflationary pressure of our time lies in a reform of the international monetary system capable of bringing the intensity of the speculative factor back to that prior to 1971, when, in spite of all sorts of problems, wholesale prices of all commodities in the industrial countries were increasing very slowly or, in certain years, were even slowly falling. The dollar, as a reserve currency and as an accumulator of value, was working relatively well; at present, it has ceased to be a reliable unit. We have to find a substitute. It seems difficult to imagine a relatively stable and reliable international monetary unit completely dependent on the discretionary policies of internal and international monetary authorities. If we want to have such a unit, those policies should be in some way conditioned by an objective standard; and it is difficult to imagine an objective standard different from gold.

However, the conditions for the working of the gold standard are bygones forever. Properly speaking, that monetary system implied the convertibility of bank notes into gold coins of a given weight for everyone — banks, companies, individuals. As for the gold exchange standard, historically we have known at least two varieties of such a system. The first one — adopted in several countries after World War I — implied the gold standard in at least one country; in the other countries the bank notes were convertible into bank notes and into credit instruments of the country with the proper gold standard. In the countries adopting the gold exchange standard, the convertibility into gold coins was not only indirect, but also restricted to companies and banks — in practice, individuals were excluded; under these conditions, gold did not circulate and, therefore, strictly speaking was no more money, but was only a medium performing certain monetary functions. It has been said that the Bretton Woods agreements reintroduced the gold exchange standard after World War II. This is true only if it is specified that we have here a second variety of gold exchange standard, since the leading monetary unit — the dollar — was convertible into gold bullions, not into gold coins, and was not open to all, but only to central banks and at particular conditions.

It is probably now worth discussing the possibility of establishing a third variety of gold exchange standard, in which the convertibility of the monetary unit — be it the ECU or the dollar or another unit — should be 1) restricted to central banks, 2) admitted only at given intervals, and 3) introduced gradually after creating a forward line of defence based on the strong currencies. Of course, the new arrangement should re-introduce the system of fixed exchange rates[1]. In short, to combat the world inflationary process a new Bretton Woods is necessary. (It is worthwhile to re-examine very carefully the plan presented by Keynes at Bretton Woods.)

5. The World Inflation and the Slowing Down in the Process of Growth

If it is true that the origin of the worldwide inflation is the one that I have just mentioned, the management of the quantity of money in the individual countries cannot be the solution. In particular, monetary policies, even if very restrictive, are bound to have effects which are limited in terms of inflation and, at the same time, costly in terms of production and employment.

Granted that there is a link between the variations in the quantity of money and those of total demand, we can accept the proposition that variations of total demand bring with them variations, in the same direction, of prices *if* the demand flexibility of prices and earnings is the rule, as was

[1] Cf. R. Higonnet, *Keynes and the Gold Exchange Standard*, Discussion Paper no. 1, International Banking Center and Department of Economics, Florida International University, Miami, November 1981.

the case in the past century. But we cannot accept that proposition in an economy such as that of the industrialized countries at the present time, where the double dichotomy referred to above has arisen and taken root. In such an economy (in which the sector of prices and earnings which do not react, or react only slightly, to variations in demand constitute the main share in terms of income and employment), a restrictive monetary policy will have as its main consequence the reduction of investment and employment; the downward thrust on the rate of increase in earnings and in prices cannot but be very limited. This, in my opinion, is the main reason why the social costs of a restrictive monetary policy prove so high in modern industrialized economies, and the results so slight as far as the inflationary process is concerned. Instead, the results are remarkable when the aim of that kind of policy is the reduction of a foreign deficit, precisely because a restrictive monetary policy strongly affects the level of activity.

In the past century the picture was different. A restrictive monetary policy, which was usually adopted by raising the rate of discount, was followed by considerable and quick results in reducing a foreign deficit (which, in gold standard, implied an outflow of gold coins). And these results were obtained not only (as is now the case) by affecting total demand and therefore the volume of imports, but also by reducing prices, thus fostering the international competitiveness of the economy, since, as a rule, prices were flexible in relation to demand. In other words, a reduction of prices — and not simply of their rate of increase — was obtained as a by-product of that policy. In fact, a policy specifically intended to combat inflation was exceptional, since inflation itself was a rare occurrence due to events external to the economic system, like a war. Today, particularly in the last ten years, inflation has become a normal state of affairs. This does not mean that inflation is now to be regarded as a physiological phenomenon. It is pathological, as in the past; and its most damaging consequences become evident as time goes on. The main damage lies in the serious slowing down of the rate of growth in the industrialized countries considered as a whole, due to a process of stop-and-go at the world level.

In fact, today every worldwide economic expansion tends to bring with it a remarkable rise in the raw materials prices; in turn, this rise tends to determine or to aggravate a deficit in the balance of payments of the industrialized countries. To reduce this deficit the governments of those countries will adopt restrictive credit or fiscal policies. As a consequence, a worldwide recession becomes inevitable. Let us not forget that, after the slump of 1975, the recovery of the industrialized economies was strong but short-lived (see Figure 1) and that a new recession took place only three years later, that is, in 1978. In 1979 there was a new recovery, followed very soon by a new recession. Each recovery was accompanied by a remarkable rise in the price of raw materials (much higher than before 1971) and by increasing deficits in the balance of payments in the majority of industrialized countries. Each recession, like the one in which we are living, was accompanied by a fall in raw materials prices. It appears, however, that the rise is greater than the fall; it appears, too, that this fall can determine a reduction in the rate of increase in industrial prices and in the cost of living, but is unable to determine an absolute fall in these prices, because, among other things, money wages continue to increase more than productivity. The overall results of these variations are inflation (which proceeds without interruption, though at a varying speed) and a slowing down in the process of growth.

CHAPTER 17

MEDIUM AND LONG TERM STRUCTURAL ASPECTS: CONCLUSIONS AND EVALUATIONS[1]

PIERRE LANGUETIN

1. Gold as a Commodity and as Money

The contributions presented under the heading "Medium and long term structural aspects for gold" have been most informative and enlightening in respect of the fundamentals of the gold market as well as a preparation for the discussion on the role of gold as a money. The analysis of market developments in recent years and the discussion of prospects for the coming decades brought out quite clearly what are the characteristics of gold as a commodity against the background of its historical monetary function. To an extent unmatched by any other commodity, the medium and long term trend of gold has been and will be influenced by psychological factors, i.e. the general faith in the future of this metal that is still prevailing in the public mind, its use as a hedge against uncertainty — political and economic — and inflationary expectations.

2. Supply and Demand

It follows from this that the medium to long term trend of gold mining can be asserted with a reasonable degree of probability and error. The same cannot be said about the total supply available to the market, since the export of gold from some of the producing countries will be governed by the

---

[1] The conclusions and the evaluations of the Session on Medium and Long Term Structural Aspects of the World Conference on Gold were drawn, at the end of the Conference, by Pierre Languetin, Chairman of the Session, by Louise du Boulay, opening speaker of the Session and representative of the gold sector's point of view and by Robert Z. Aliber, representative of an academic point of view.

availability of other export receipts, gold being kept as a means to finance any remaining balance of payments deficits. Predictability of gold supply will also be affected by disinvestment on the part of private or official holders at times when prices appear to have reached a peak or monetary authorities are in need of foreign currencies.

On the demand side of the market, it can be taken pretty safely that physical demand for jewellery fabrication and industrial uses will develop in line with economic growth. Jewellery demand, however, may vary according to fashion and decorative needs and gold can be replaced partly for industrial uses by other metals or substances.

Investment demand is undoubtedly the less predictable of all the factors affecting the gold market. As an investment instrument, gold only qualifies in the long run if returns — in the form of higher prices — are assumed to reach at least comparable levels to those obtained by other investment means.

3. Before and After 1970

The situation that exists since the early seventies is obviously in sharp contrast to the one that prevailed at times of gold or gold exchange standard. Factors affecting the market had been then predominantly determined by monetary demand of central banks.

This, at least, lasted for as long as industrial and investment demand did not exert upward pressures on the price, a phenomenon that eventually led to the creation of a two tier market. From then on, it became overwhelmingly clear that the price of gold as a commodity — which must be allowed to reflect the state of the market — was not compatible with the price of gold as a money, which, by definition, must be a stable price.

Since the beginning of the seventies, it can be argued that the price of gold as a commodity will be the determining factor, the current price of gold being doomed to oscillate around that level. As one of the speakers remarked: "Large-scale, short term buying of gold, when added to commodity demand and long term investment purchases, can force the price of gold to rise rapidly well above a lasting equilibrium price. But as the price rises, commodity usage declines... the same mechanism works in reverse."

However, the price of gold will also be influenced by the interest rates, rising when the latter tend to decline, declining when the interest rates are high, as has been the case during the last years.

This observation carries, in my view, certain implications. Recent experience has shown that the short term fluctuations of the gold price can be very sharp and rapid as a consequence of speculative reactions or on the spur of unexpected events, particularly in the political field. Over a certain period, however, it could be argued that these fluctuations will tend to remain within certain limits. The reason for this is that inflation and interest rates tend to evolve hand in hand. If inflationary developments and expectations tend to push up the price of gold as a commodity, the admittedly parallel rise in interest rates will tend to put a limit on the price of gold or even reverse the evolution. In such a case, gold would imperfectly play its alleged function as a hedge against inflation.

4. Two Final Observations

The preceding remarks — partly of a theoretical nature, partly based on practical experience — suggest two final observations.

In the first place, gold by itself does not qualify for a sound and stable monetary standard. In the past, stability could only be achieved because of the central banks' demand for unlimited quantities at artificially high prices. Stability was defeated when market prices rose above monetary price.

In the second place, gold as a commodity is sure to retain its importance. It is most likely to support a profitable activity for mining industries, dealers, jewellers and other utilizers.

Investors, however, will possibly have to be more sophisticated than in the past, having to weight the profitability of gold against that of other investment means, and to compare inflationary expectations with interest rate developments. Fuller awareness of the gold market characteristics — of its medium to long term trends — could help towards better efficiency and stability. May the Conference on gold have contributed to this positive result.

LOUISE DU BOULAY

It is my task this morning to try to summarize the views of our session on the structural and long term aspects for gold. I intend to do this by using the framework of my own paper and working in comments of the earlier speakers from our session. Prof. Aliber is going to cover the later ones. I fear that in so doing, I shall probably fail to do justice to the papers of the other speakers, but Mr. Languetin and I reckoned that most of us came to roughly the same conclusions and this would therefore be the least repetitious method of approach.

5. The Supply Aspects of the Market

First I would like to cover the supply aspects of the market. The 1970's were characterized by sharply rising prices and indeed highly volatile prices.

However, despite this rise there was, in fact, a contraction in the mine supply of gold by something in the region of 25 percent. This was explained by Tom Main in his paper and was largely the result of a deliberate policy on the part of the mines to extract lower grade ores as it became economic for them to do so.

South Africa has long been the dominant producer of gold and still accounts for some 70 percent of production. However, while production in South Africa, North America and indeed Australia has been declining since 1970, the decline has to some extent been offset by increases in other countries, particularly in Latin America, the Philippines and Papua New Guinea.

I think it is worth saying that the mining industry regarded the 1970's with somewhat mixed feelings. On the one hand, the removal of official price restraints gave the mining industry in general a new lease on life, but at the same time, as Tom Main has pointed out, the rise in the inflation rate of the 1970's — which to a large extent caused the boost in the gold price — created a whole new set of difficulties for the mining industry. Both working and capital costs rocketed and uncertainties about the future made planning extremely difficult. Nevertheless, the boom of 1979 and 1980 caused a positive surge in exploration and new development and, as Mr. Hanselmann noted, as a result of our research into these activities, we are anticipating some increase in production to 1985.

In summary, we, supported by Tom Main, are assuming that South African production will remain fairly static to 1985, more or less regardless of the price. This is due to the inbuilt conservatism of the mining industry in South Africa, which as a result of both the depth and the size of mines, has to plan with great care and has also — I would hasten to add — to have great confidence in the future before committing itself to the enormous capital investments required for new mine developments.

In other parts of the world, the enormous number of small projects which came to light in 1979 and 1980 — some of which were in fact on the books before that — seem due to open up by 1985 and could result in a substantial increase in production in other parts of the world. Some of these projects may fail to start because of the recent fall in prices but for the most part we think that those with low production costs will open. In Brazil and Latin America, where production tends to come from alluvial sources requiring short lead time and little capital expenditure, we assume that production will tend to be more price elastic than in the countries such as South Africa.

Our conclusion is therefore that while South Africa's share of production will tend to decrease because production will remain static, this will largely be compensated for by increases in other parts of the world to 1985. From then onwards as South African production tends to decrease, the increases in the other parts of the world will be less able to compensate for this. Towards the end of the century we are therefore looking for a reduction in total mine supply.

One final comment on this subject is that we believe that the higher cost of new capacity coming in — which Tom Main gave us as $350-400/ounce in

South Africa, and which our own researches in other parts of the world would tend to support — will place a rising floor under future gold prices.

The total supply of gold tended to increase through the 1970's, despite the fall in mine production, due to availability from two other sources, namely sales from the Eastern Bloc and sales from the official sector.

Both Russia and China have long histories of involvement in the gold markets and both are also producers. The Russians have been the chief source of Communist gold to the market over the years and in substantial quantities. We assume that Russia sells gold in direct response to foreign exchange requirements and the quantities that are sold have therefore varied in relation to her requirements. It is difficult to predict Russian behaviour because of the secrecy that tends to shroud all her activities, but we believe that some general points have implications for the future.

We, at Gold Fields, estimate that Russia produces some 300 tons of gold annually and having taken account of her domestic requirements, some 200-250 tons are therefore available for sale to the market each year. At the same time, there are growing signs that the Soviet system is incapable of coping with the complex modern economic conditions that we face today. There have been articles in the press recently which highlight the general level of incompetence and inefficiency which seem to affect all aspects of production in Russia, whether in the mining industry or in the agricultural sector. We can only assume that this situation will continue and result in a worsening of the balance of payments situation in the future. We assume that gold will continue to be a major source of foreign exchange revenue to Russia because it is one of the few goods that she produces which is acceptable to the West and which can be sold quickly and with relative anonymity. At the same time, there have been a number of estimates that suggest that Russian oil exports, which have been a very substantial foreign exchange earner to her in the past, are declining and it can be assumed that her dependence on gold will, if anything, increase. I would therefore conclude that Russia is likely to continue to sell a minimum of 200 or 250 tons per annum to the market for some time to come.

A great deal has been said about the official sector and the central banks' use of gold, a summary of which will follow this session. I intend to concentrate mainly on the factors which will affect the physical supply and demand for gold to the market in the future.

During the late 1960's, and for most of the 1970's, the official sector was a source of very substantial quantities of gold to the free market. We believe that 1980 marked a turning point in this respect and, in fact, marked a return to the pre-1965 situation in which the central banks were the main buyers of the gold coming on to the market from the mines each year. In the future, it seems probable that the official sector will continue to be a net purchaser of gold from the market. There are a number of reasons for this belief.

Any discussion of the likekihood of a return to the gold standard is really not the province of this session, but I must admit that from what I have heard in the last few days, I am confirmed in my belief that it is very unlikely. We therefore assume that for reasons of political reality, the present laissez-faire system, which now operates in the official gold market, will remain. Central banks will continue to be free to hold, trade, or price their gold as they see fit. There seem to be a growing number of central banks who are in fact interested in utilizing their existing gold reserves in various ways and, in some cases, increasing them. The sales of the late 1970's, by the IMF and US Treasury are now seen to have been a mistake, given the price at which they took place, though I do take Mr. Wittich's[1] point that at the time of the IMF sales, all those concerned in fact benefited in some way or another. At any rate, there are now few who are in a hurry to sell their gold holdings, except in cases of extreme financial distress.

While sales look unlikely, purchases, on the other hand, seem more likely, certainly in the case of countries with a low gold content in their reserves, such as the OPEC countries and the recently industrialized countries. In comparison with the industrial countries and, funnily enough, some of the other developing countries, the portion of gold that these other countries hold in their reserves is remarkably low and there are indications that they wish to redress this balance. It is also worth saying that gold remains the only truly international and therefore probably apolitical reserve asset available and is valued for this alone.

From these facts we conclude that the official sector will no longer be a supplier of gold to the market and will normally be a taker of gold from the market. We also believe that this could bring some stability to the market and, perhaps along with rising mining costs, provide some form of base support to the price. It seems less likely that the central banks will be selling at the peaks as their holdings are far from speculative.

6. The Demand Aspects of the Market

I would like to turn now to the demand side. Demand for gold can be divided into two areas. The first is the physical demand for jewellery fabrication and industrial uses, and the second is investment demand.

Trends in jewellery demand are really rather difficult to tie to any single factor as the motives for buying it vary enormously from one country to another. In Europe, jewellery has been bought traditionally as a status symbol and as a means of decoration and, Mr. Torboli pointed out, often to mark some special occasion such as a christening. In the Middle East and Far East, on the other hand, it is very difficult to distinguish between the buying of gold for decorative purposes and the buying of gold for investment. I think this difference can be most readily appreciated if it is realized that in

---

[1] See Gunter Wittich, *The Role of Gold in the International Monetary System*, Part III, Chapter 20.

Europe and in the United States, we tend to buy gold which can often be of a low carat and at a very high mark-up to the gold content, while in the Middle and Ear East, it is seldom that people buy jewellery of less than 22 carat, which is sold on a very low mark-up. The holding of gold jewellery in these areas is therefore much more fluid and we have observed a tendency, over the last 10-12 years, when prices rise, for these holders of jewellery to cash it in. Nevertheless, they do seem to return to the market when prices are lower and buy back again.

However, although jewellery sales do appear to have an inverse relationship to price movements, the experience of the 1970's suggests that eventually even buyers in the industrialized countries do adjust to higher prices in time and we have no reason to suppose that this will change in the future. We are therefore fairly optimistic that jewellery sales will show some growth in the long term.

In the UK we have now obtained some statistical evidence to show that jewellery sales are closely related to levels of real disposable income. I believe that Intergold in the US have come up with the same conclusions. Jewellery sales will therefore benefit in periods of economic growth and, despite the current gloom, I am sure that we can anticipate further growth in the industrialized world. Meanwhile, the price-sensitive demand in the developing countries will also continue, in reverse relationship to movements in the gold price.

We have also observed signs of a growing awareness of the unsatisfied demand for higher carat jewellery at prices which are more closely related to the gold price in the industrialized countries, more on the basis of sales of jewellery in the Middle and the Far East. As a mining company, this will obviously suit us very well because it means that for every item of jewellery that is sold, there will be a higher gold content. We also hope that it will encourage people to buy jewellery in the belief that it is indeed a good investment.

Here, I would like to make further mention of Mr. Torboli's speech, which described the Italian jewellery industry in great detail. I think it shows how much one country alone can achieve as a jewellery manufacturer and the possibilities that can exist both in domestic and export markets. I am sure that most of you are aware that Italy is the world's largest producer of gold jewellery, accounting for 50 percent of total jewellery production in Europe and 25 percent in the free world — no mean achievement. He also pointed out that the Italian jewellery industry, despite the development of technical skills and high quality designs, is facing difficulties, as does the mining industry, due not only to rising and volatile gold prices and rising costs, but also due to competition from developing countries, where the cheap labour is obviously the main factor.

I can only say that during a recent visit to the Middle East, I was surprised to see the amount of Italian imported jewellery and to hear the comment of some of the local manufacturers who were complaining that they were unable to manufacture jewellery locally for prices that were as competitive as those of Italian imports. So, you may have problems in the Italian jewellery industry, but I can assure you that they do not seem to be as bad as those in some other parts of the world.

We are not so optimistic about the long term future of other forms of industrial consumption, namely electronic, dental, general plating, etc., though I must point out that these are small in comparison to the use of gold for jewellery. As a result of the high gold prices, companies have expended a great deal of time, effort and money into researching methods for reducing the quantity of gold used. Furthermore, these efforts don't cease when prices drop. In the long term, we are therefore looking for some measure of decline in industrial uses, for although there is likely to be growth in the electronics industry in particular, it is unlikely to be matched by growth in gold off-take.

This brings me on to investment demand which can be divided into two sectors: the physical demand for coins and small bars, and large scale investment demand.

It is interesting to note that sales of bullion coins have increased at a really remarkable rate since 1973 due in part, as Herbert Coyne pointed out, to greater availability and undoubtedly also to advertising. In addition to this, in the last year, we have become more aware of the rapidly growing demand for small bars, to which Tim Green drew attention in his paper.

This demand for coins and small bars can really, I think, be explained by a mixture of motives, both investment and decorative. It is possible that people are looking for a better value-for-money method of buying gold and this may, to some extent, be reducing normal jewellery demand because many of the smaller coins and bars are worn as jewellery. Perhaps it is too early to say and it is possible that the demand for small bars is a gimmick or new fashion. It is also very noticeable that within Europe the preference for either coins or bars depends very largely on the tax situation that prevails in each country. I think, too, that as Herb Coyne suggested, the availability of gold in small denominations which are affordable by everyone, has boosted small scale investment demand so that it is no longer the preserve of the rich. Most people can now reach an outlet where they can find a form of gold investment tailored to their needs, whether physical or in the form of certificates, or the gold pass-books developed by some banks in New York, Switzerland and Germany. At any rate, it is encouraging that sales of coins and small bars held up so well last year, despite the depressed price outlook, showing that there can be a good basic demand, both in periods of falling as well as rising prices.

Investment demand for gold in general would appear to be dominated by inflationary expectations and political apprehensions. Both are somewhat hard to predict and can have an effect on price movements, both upwards and downwards. In some ways, I can't help feeling that the subject of investment in gold is surrounded by a certain amount of myth which often does not bear close examination and the results of which can be somewhat unexpected. I would therefore like to briefly talk about a few of the subjects which were mentioned in relation to investment demand for gold.

Past observations and experiences in many countries do support the idea that gold remains an invaluable asset in areas of political and indeed economic instability or upheaval. Gold has played a role throughout history and I see no reason to expect this to change, given its portability and universal acceptability.

A number of speakers have suggested that inflation is probably the major influence on the gold price and I don't think that I would disagree with this. It would also follow that if inflation could be defeated then gold would lose some attraction as an investment. Whether it is fortunate or unfortunate from the gold producers' point of view, I don't believe that inflation will be defeated in the longer term as it seems endemic to our system and I therefore assume that gold will continue to be acquired as a hedge against it.

A second question arises — whether gold actually has been a hedge against past inflation. Professor Jastram has demonstrated most convincingly in his book *The Golden Constant* that during the era of gold standards and fixed prices, gold, far from being an inflation hedge, remained constant, while, as would be expected according to the operation of the system, prices automatically fluctuated around that constant gold price. However, since we no longer have gold standards nor fixed prices, the situation has changed and in the last ten years the gold mine has more than outperformed the rate of inflation. Perhaps it is safe to say that gold has now become an inflation hedge, somewhat belatedly supporting assumptions made about the past.

Professor Van Tassel mentioned the adverse effects of high interest rates on gold investment, but I must admit that I tend to agree here with Professor Aliber, who said that if you try to correlate short term movements in gold price with inflation rates, there is no observable relationship. I think it is probably fair to say that interest rates should be more realistically regarded as alternative investments to gold. The carrying costs of holding gold when interest rates are high are perhaps a problem, but more to the point, investors tend to buy interest-bearing assets in preference to gold, when the expected rate of return is greater.

The final point that I would like to make on the investment side, as stressed by Mr. Hanselmann and Mr. Coyne, is that the world's investment climate is now much more favourable to gold than was the case in the past. It is really only very recently, in fact in the mid-1970's, that gold has become legally tenable by the citizens of a number of countries in the world, including the US, the UK and Japan. The developments in Japan of the last year show the potential that can exist in new markets. And, I think, as Herb Coyne pointed out, such movements tend to feed on one another and the example observed in one country can lead to developments to a further stage in other countries.

This tendency towards growing investment in a wider geographical area and by a wider number of people will, like jewellery sales, be helped by the greater availability of gold in a number of forms and by the development of new openings, whether they be banks, brokers, investment institutions or whatever.

It must also be remembered that gold can only qualify as an investment instrument in the long term if the returns, namely in the form of higher prices, are at least comparable with those of other investments. I see no reason to believe that this will not be the case in the future. As Herb Coyne suggested, people tend to regard gold as a faded and unattractive asset at the moment, because the price has fallen so sharply. I feel, as does he, that investment in gold is cyclical, that we are at present in a cyclical downturn and not seeing any fundamental change in the attitude to gold. When circumstances alter and when the attraction of other investments decrease relative to gold, then gold will once again come back into vogue.

7. Final Comments: Objective Analyses and Professional Interests

I would like to reiterate the conclusion drawn as to the long term structure of the gold industry by the speakers in our session.

Basically, we are anticipating that static mine supply and the absence of official sales will result in less gold being available to the market than was the case in the 1970's, despite the possibility of a high level of sales from the Eastern Bloc. Meanwhile, as I have said, we are optimistic that consumption of gold for jewellery and, indeed, for investment purposes in both the private and the official sectors will tend to increase, given that there will obviously be fluctuations in line with price levels. The net result will be a tendency towards a shortfall between the supply and demand for gold in any one year, and that shortfall will have to be met by disinvestment. In the past it has been very obvious that in order to encourage disinvestment, prices have had to rise.

We also feel that volatile prices are, in the long term, no help either to the mining industry or to the jewellery industry. There are signs that perhaps in the long term, prices will be less volatile than in the last few years. It is very unlikely that short term price movements will be less volatile, largely due to the pressure of external circumstances which can result in a combination of high fabrication and speculative demand at the same moment. However as Professor Van Tassel pointed out, the gold market has built-in automatic mechanisms for bringing about reversals in price trends, in either direction. If prices rise too sharply fabrication will drop abruptly, and it is likely that disinvestment will then occur, causing the price to fall and leading to a pickup in fabrication, starting the whole cycle all over again. By the same token, the basic commodity and monetary influences on the gold price will at times be influnced by external circumstances; rising inflation, political upheavals, currency movements, etc.

I have one final comment. Lest as a representative of a mining company I be accused of concluding that in the future the price will rise, through vested interest, but perhaps more slowly, and I hope more slowly than in the 1979-1980 period, I would add that we, as a mining company, are putting our money where our mouth is and are continuing to expend enormous sums of money on exploration and the development of gold mining projects.

ROBERT Z. ALIBER

8. Three Questions on Gold

I believe that our session had no consensus and, indeed, I would like to indicate the lack of a consensus by asking the audience to participate in a quiz or poll. Three questions will be asked.

The first question is: how many individuals in the audience would like to see a role for gold as a money in the international system or in the domestic system? The second question is: how many would like to see a future role of gold as a commodity held by individuals for possible increases in its price? Now the third question is how many responded positively to both questions? Those in the audience who answered yes to both questions have a real problem, which was the problem of our session. For we must decide whether we want gold as a money or whether gold will be held because it promises a competitive rate of return. The inconsistency is that if gold is going to be used as a money, its nominal rate of return — as Professor Jastram has so eloquently written — will be zero. So the rate of return on gold will be lower than the rate of return on other assets.

If instead gold is going to be used as a commodity like copper and silver, it will not be a better inflation hedge than any other commodity.

In the course of a year or two, I attend numerous conferences at which representatives from the petroleum industry and the copper industry and the tin industry speak. Many of them make statements, like the statements of the representative of the gold industry, indicating that the real price of "their product" will rise more rapidly than the prices of most other products. These statements are unlikely to be true! That literally cannot be true!

In our session, the economists, Professor Jastram, Professor Sylos-Labini and I wanted to maintain a primary role for gold as a money, although we differed on how extensive or comprehensive that role would be. The representatives from the industry instead wanted to see a role for gold as a commodity.

9. The Central Issue: Gold and the Monetary Environment

The central issue involves the relationship between gold and the monetary environment. The reason that we are at this Conference — as Anna Schwartz[2] reminded us yesterday — is that we have lived through a very unstable decade. Monetary growth has been quite unstable. In retrospect, monetary growth during the nineteenth century appears more stable — at least at a casual or superficial level. Most of us forget, especially in the United States, that the long run price stability resulted from the frequent financial collapses under the gold standard.

The previous speaker raised a central question: what is the outlook for inflation? Meaningful forecasts about the future demand for gold require statements about changes in the world price level. The demand for gold is especially sensitive to changes in the anticipated rate of inflation.

Let me therefore speak briefly on three different views about the US inflation rate, since the United States has such great importance in determining the world inflation rate. While one or two other countries may have lower inflation rates than the United States, by and large, the world inflation rate is set by the US inflation rate because of the large size of the US economy in the world economy.

The consensus forecast about the US inflation is about eight percent for 1982 and 1983. A second forecast can be inferred from the financial markets. The forecast of people who are in the bond markets is for an inflation rate of eleven or twelve percent; this forecast is inferred from money interest rates.

The third forecast is my own. I am not a professional forecaster, so perhaps this forecast should not be taken too seriously. It has been a decade since the inflation outlook has looked as favourable in the United States. The inflation rate in producer goods prices, and by and large all the news, look good. The oil market price news looks unbelievably good. The food market news looks unbelievably good. There is a considerable amount of wage re-negotiation in autos and steel, airlines, and even in trucking. My forecast is that the inflation rate at the end of 1982 will be about five percent and that the inflation rate will continue to decline modestly. Perhaps this outlook is too optimistic. One reason for an inflation cut is that the government benefits from the inflation tax, in that the rate of interest that the government pays on its debt declines because the consumer price level is increasing more rapidly than the interest rate. That free ride is now over in the United States; indeed the real interest rates now are so high that they are really a major cause of President Reagan's budgetary deficit.

---

[2] See Anna Schwartz, *The Past, Current and Prospective Role of Gold in the US Monetary System*, Part III, Chapter 21.

If my own inflation outlook proves correct, and the inflation rate declines more rapidly than the consensus forecast, the future prospects for an increase in the gold price are really very very modest. Indeed, the current gold price contains an implicit inflation forecast of ten or eleven percent. So if my forecast is correct, there is likely to be a very sharp decline in the price of gold.

The significant point is that we cannot make meaningful statements about changes in the market price of gold — statements about the demand side and the supply side — without indicating the price level environment in which those forecasts are going to be operative.

10. No Consensus between Economists and Gold Companies

So there was no consensus in our session between the economists who favour a monetary role for gold and the representatives from the gold companies. I think the economists generally recognized the importance of significant decline in the inflation rate before gold will once more have a significant monetary role.

Thus, even though central banks may buy gold in the current environment, central banks are not likely to sell gold until there is reasonable stability in the world price level and in the market price of gold. And the reason is the political embarrassment associated with selling gold if there is any likelihood the gold price will then increase. Central banks will finance their payments deficits by some other ways. So it is very unlikely that there will be a new role for gold in the system until the inflation rate falls, and the market price of gold appears to be stable.

Part III

NATIONAL AND SUPRANATIONAL MONETARY AUTHORITIES' POSITION

Session chaired by René Larre

CHAPTER 18

POSSIBLE AND DESIRABLE ROLES FOR GOLD IN THE INTERNATIONAL MONETARY SYSTEM: RECONSIDERATION AND PROPOSALS

PETER M. OPPENHEIMER*

1. The Historical Background

The monetary role of gold has been fairly steadily shrinking for the past 150 years or so. The basic reason for this is that bank money or credit money turns out, on balance and for most purposes, to be a superior instrument. Some of the very qualities which made gold the best form of money in an earlier age have caused it to be superseded in the industrial era, particularly its natural scarcity and intrinsic value. Gold has been *too* scarce, too costly to dig up, and too valuable to serve any longer as a general medium of exchange — though by the same token it retains many attractions as a store of value. Bank money comes much closer to being a "pure" medium of exchange, like a catalyst in a chemical reaction. One aspect of this is that, depending on the precise institutional arrangements in operation, it can yield interest to its holder. Gold is, of course, liable to yield real capital gains over time, but the extent and timing of these are uncertain, so that they do not serve to enhance gold's attractions as a medium of exchange. On the contrary the operation of Gresham's Law and the periodic dominance of the gold market by speculative dealings tend to make gold unusable in this respect.

1.1. When bank money first developed, gold commonly retained the function of monetary base, albeit with only a lose linkage to the total quantity of money. To maintain bank liabilities convertible on demand into gold or other metals prevented persistent over-expansion of the money

---

* In writing this paper I have drawn on my earlier paper presented to the Bank of England Panel of Academic Consultants, "International Monetary Arrangements: the Limits to Planning" (Bank of England, 1979).

stock. It did not necessarily prevent over-expansion in the short run; if this occurred it was liable to be corrected by way of bank failures and crises. Such crises might also be precipitated by the commercial failure of a single bank, even in the absence of any general over-issue of bank money. Market information was not reliable enough to be able to distinguish between these two types of bank failure; and indeed the two might occur jointly. The felt need to enhance the short run stability of the banking system by forestalling such crises gave the main spur to the emergence of central banking in the nineteenth century — which put a further series of nails into gold's monetary coffin.

1.2. At first central banks saw themselves as the handmaiden rather than the executioner of gold money. Their function — typified by the Bank of England up to 1914 and even up to 1931 — was not to supersede gold as determinant of the monetary base but only to surpervise it and iron out short term difficulties which might otherwise upset the system. Admittedly, as Robert Triffin has pointed out, the belief that money stocks based on gold were exogenous variables to which national price levels adjusted was an illusion. The variable which adjusted to keep the system going was not so much the price level as the relationship between gold stocks and national money supplies. As the use of bank deposits grew, monetary gold was increasingly withdrawn from circulation and concentrated at central banks. However, this was largely a spontaneous evolution, and not deliberately fostered by the authorities (though they may have fostered it indirectly by their contribution to the stability of the banking sector).

1.3. In the twentieth century, with the emergence of active monetary management and the expansion of the economic functions of government, the balance of power between gold and central banks was shifted clearly and decisively in favour of the latter. Gold was *not* to act as the effective constraint on the creation of money by national authorities. This view was implicit in the decision of the Genoa Conference of 1922, advocating extension of the gold exchange standard to relieve a shortage of gold. At that time the idea of changing the currency price of gold (other than allowing it to fluctuate as a temporary measure in wartime or other emergency) was not yet regarded as respectable. This was changed — or so one thought — by the experience of the 1930's. The collapse of the restored gold standard after 1931 led, by way of floating exchange rates, to an endogenously generated increase in the value of world gold stocks. As has frequently been emphasized, the relative prices of the major currencies were not very different after 1936 from what they had been in 1930. The main change was the more or less uniform increase in the currency price of gold. But it had been achieved in a disorderly way.

1.4. The IMF Articles of Agreement of 1944 were an attempt to prevent such disorder in future. Along with an insistence on free trade and current payments, they provided for adjustments in the currency price of gold, either through an agreed uniform change in par values or alternatively through a series of par value changes undertaken on account of successive "fundamental disequilibriá" in national balances of payments. (It will be recalled that all par values were denominated, directly or indirectly, in gold.) The Articles did not attempt to spell out the exact circumstances in which par value changes would be appropriate, though it was generally understood that domestic financial policies should not be subordinated to the requirements of external as against internal equilibrium. Nor did they attempt to spell out just what was needed by way of reserves and credit facilities to make the pegged-rate system viable. However, along with traditional reserves in the form of gold and currencies, they did provide for the new conditional lending mechanism of IMF drawing rights, to assist countries in coping with particular payment deficits. Such lending was a modern international analogue to the nineteenth-century domestic functions of central banks, providing last-resort credit in order to allow a gold-based system to operate in a reasonably crisis-free manner. The IMF Articles thus left many questions unanswered; but they were operationally realistic.

2. Gold, Reserve Currencies and International Payments Equilibrium

The orderly change in the gold price envisaged in the IMF Articles was never brought about, even though it was evidently needed by the later 1960's (or indeed earlier). In practice it fell to the US authorities to decide, and they refused to seek an alteration in the effective par value of the dollar. (The two changes of par value passed by Congress in 1972 and 1973 were formal and of no economic significance.) To attempt a full explanation of US policy towards the dollar is beyond the scope of this paper. But one factor, or set of factors, was a misperception of the position which the monetary role of gold had reached. Crudely speaking, many economists mentally extrapolated the extinction of gold's role in domestic exchange to the international monetary system: having removed gold from the domestic circulation, banking tecniques and monetary management could and should now do the same in the international sphere. Not to do so seemed on this view an affront to reason, a sentimental clinging to outmoded institutions.

2.1. Obviously it is possible nowadays to avoid having gold at the centre of international monetary arrangements: the world has been doing so since August 1971. The point, however, is to understand what difference this makes to the monetary system, and not simply to assume that it need make no difference at all. Gold has continued to matter in the international sphere long after its disappearance from domestic circulation, because there is no global central bank or government. International monetary arrangements are made between national authorities, each of which is sovereign in its own domain and must serve national political goals. Only very limited relinquishing of national monetary sovereignty to international bodies such as the IMF is possible. At the same time, there is no way in which round-table cooperation among national authorities can determine in some exogenous way the volume and composition of international reserves. On the contrary, any viable monetary scheme must allow the international system to generate endogenously, through the pursuit of equilibrating policies by national governments, an appropriate distribution, rate of increase and hence volume of reserves, and not seek to impose limiting values on these variables in advance. In other words, market forces must be brought in to help reconcile conflicting viewpoints and preferences, not merely at the level of individual traders but at government level too. One may argue about whether this is a strictly theoretical finding as well as an empirical one.* It certainly has something to do with the uncertainty of human affairs and the need to respond to unforeseen contingencies, as well as with the facts of national sovereignty and national monetary management.

2.2 Of course, in order to ensure that equilibrating policies are indeed pursued by national governments, finance for disequilibria may need to be arbitrarily limited. This applies vis-à-vis any single country. But for the system as a whole the relationship is the other way round. The proper rate of increase of reserves will depend on the prevailing equilibrium or "normal" structure of international payments, and cannot be used as a lever to determine the latter. (Even for those unable to discern this in the 1960's, the development of global reserves and payments patterns in the 1970's must constitute persuasive evidence.)

2.3. In this context, the manner in which different types of reserve asset are supplied and managed turns out to be inter-related in specific ways with the exchange-rate regime. On the one hand, exchange rates may be left to float. In that case the markets called in to reconcile diverse national preferences are simply the foreign currency markets. The price of gold in any single currency will fluctuate with that currency's exchange rate — although, if there is no worldwide inflation, the weighted average price of gold will be more nearly stable. On the other hand, suppose that it is desired to fix currency exchange rates permanently or until further notice. How can this be achieved? What is to be the numéraire and fulcrum of the system?

2.4. In the past 150 years worldwide fixed or pegged exchange rates

* An economist being, after all, someone who, on seeing anything working in practice, asks whether it could possibly work in theory.

prevailed (a) from 1870 (or thereabouts) to 1914 (plus the brief restoration of 1926-31) and (b) from 1945 to 1971. In both periods, gold functioned not merely as one reserve asset among several, but as global numéraire and basic reserve asset. In addition, both periods saw a dominant, "hegemonic" currency — sterling before 1914, the dollar after 1945. This was lacking in the inter-war years. Many economists have been inclined to dismiss the significance of gold in this story, and to focus on the dominant currency as the crucial element. It is likely that a single dominant currency facilitates the maintenance of a fixed rate system, because such a currency will generally be less subject to speculative disturbances than two or more roughly equipollent currencies. This, however, is a crude assessment, and can in principle be modified by reference to modern portfolio theory. Reserve portfolios consisting of three or four different assets whose prices show little positive covariance may be quite stable, regardless of varying prospects for the assets individually. In any case, even if the presence of a hegemonic currency is necessary for the existence of a fixed rate system, it is not sufficient. Before pegging their currencies to the central one, other countries will wish to be assured that their monetary authority will observe balance of payments discipline and keep the currency stable in terms of some outside standard or criterion.

The subservience of Bank of England monetary policy before 1914 to the state of the gold reserves was a major factor sustaining confidence in sterling, quite separate from the United Kingdom's dominant weight in international trade and investment. Analogously, the status of the dollar after 1945 was called into question when and only when its convertibility into gold was first eroded and then formally abrogated. For the reasons stated above, convertibility into an internationally managed asset such as SDRs was not and is not a feasible alternative.

2.5. To guard against misunderstanding, the scope of the foregoing argument may be spelled out somewhat further. The dollar does not require "backing" by gold in order to function as a medium of exchange or as a reserve currency. Although its dominant position among reserve currencies has been somewhat diminished, largely as a result of its weakness in the exchange markets during the 1970's, it will remain the most important reserve currency for the foreseeable future. Moreover, the use of gold as an active reserve asset depends heavily (although not entirely) upon its enjoying a stable price in terms of a few major currencies, which under the Bretton Woods system meant in terms of the dollar. But, despite these points, gold also had an independent monetary role in the Bretton Woods system, by virtue of its properties as a commodity numéraire and reserve asset. In the first place, it furnished the criterion of balance of payments discipline for the United States, as well as a key instrument for observing that discipline. These features were lost when the dollar ceased to be convertible. Secondly, gold made — in the Bretton Woods era no less than under the pre-1914 gold standard — a unique contribution to the equilibrium of the monetary system as a whole (summarized in the following paragraph). What all this amounts to is that gold enabled the dollar to act as the centre currency of a global pegged-rate system, rather than a floating-rate or currency-bloc system.

2.6. When an increase in the gold price was mooted in the 1960's, many observers regarded it, wrongly, as a method of giving the United States additional leeway to *finance* deficits rather than as a weapon of adjustment necessary to end the deficits. This misperception stemmed partly from the fact that undue attention was focussed on the instantaneous increase which a higher price would imply in the value of existing gold stocks, to the neglect of the arithmetically smaller but equally important effect on the value of new gold flows. Gold being a commodity and not a debt, its inflow to the monetary system constitutes a margin of payments surpluses unmatched by deficits. The higher the currency price set by monetary authorities (under Bretton Woods, *de facto* the US Treasury), the bigger the inflow — mainly because output is worth more per ton and private buyers take a smaller proportion of it — so the wider the absolute margin of net surpluses. A reasonable size margin of this type is necessary for the viability of a pegged-rate system, at any rate in an expanding world economy where monetary authorities will tend to be more tolerant of persistent small surpluses than of equivalent deficits. If such normal or equilibrium surpluses — for that is what they are — entail deficits elsewhere in the system, a viable structure of pegged exchange rates will be unattainable. Some currency or other will always be a devaluation candidate, and sooner or later, if the situation is allowed to persist, speculation against it will reach unmanageable proportions.

2.7. On this analysis the inflow of new monetary gold had been too small for comfort ever since World War II, as Triffin first noted. The United States became the chief victim of the situation, because it was effectively committed to meeting, in either dollars or gold, the system's residual demand for reserves, i.e. the demand generated by the sum total of other countries' policies, less any contribution from new gold. Furthermore, attempts to solve the problem by upvaluing other currencies rather than devaluing the dollar must fail, because such upvaluations would lower the average price of gold and therefore ultimately *increase* pressure on the US gold reserve, rather than reduce it. (In practice, it simply proved impossible to secure the upvaluations.) Because of the exceptional strength and confidence enjoyed by the dollar at the beginning, and policy moves later by the United States to prevent conversions of dollars into gold, the imbalance of the system was able to continue for a considerable number of years — so much so that the imbalance came to be wrongly seen as a trademark of the system itself.[1]

3. The Position of Gold under Floating Exchange Rates

Since the US authorities were unwilling to maintain the dollar's gold convertibility by adjusting its price, the dollar/gold link had to be abandoned; and without gold to hold the system together, floating proved unavoidable. Incipiently since August 1971 and overtly since March 1973 there have been floating rates among the major currencies. Coincidentally, the pattern of world payments was drastically altered by the big rise in oil prices at the end of 1973, and the subsequent structural surplus on current account of the low-absorbing OPEC countries. This has put the oil-consuming world as a whole into deficit on current account. Financing of the imbalance has been carried out mainly through the intermediation of the commercial banking system. In the process, deficit countries have on balance borrowed more than they need to finance their deficits, and have thus also added to their gross foreign exchange reserves. As far as gold is concerned, two questions now arise. First, what is, or can be, its monetary role in a world of floating exchange rates and OPEC surpluses? Secondly, should the world be looking towards eventual restoration of a pegged-rate system, either as a desideratum or as a likelihood; and how should the role of gold be viewed in this context?

3.1. The first question may be half-answered with Milton Gilbert's celebrated final sentence in the BIS Annual Report for 1973:"Down in the vaults, gold remains unused — but not unloved". Price uncertainty and volatility has lessened the usefulness of gold as an exchange reserve. Both selling and buying of gold by official agencies (and by other risk-averters) is discouraged. Large official holders may see an additional deterrent in the (possible) influence of their own transactions upon the price. If a large transactor is involved on each side of the market, bilateral price bargaining may be necessary; or it may be sidestepped by using gold reserves as collateral for an international loan (as in the Bank of Italy-Bundesbank case in the mid-1970's). The one country that was a regular seller of its gold reserves from 1975 to 1979 was the United States. The US Treasury's gold auctions were motivated partly by a wish to limit US import of the metal and partly by a wish to encourage the belief that gold was on the way out as a monetary asset.

---

[1] The foregoing analysis is basically due to Milton Gilbert. For a fuller summary of his contribution, see Appendix A.

3.2. Apart from price uncertainty, a further reason for central banks' reluctance to sell gold (though not to buy it) is that inflation — the trend of prices, as opposed to fluctuations around the trend — has enhanced its attractions as a precautionary and speculative asset. Actually, inflation has a two-edged effect, since it also adds to the price uncertainty noted in the preceding paragraph. Central banks' hoarding propensities have continued to be encouraged by the nature of gold as a commodity money ultimately independent of any national laws or monetary management. For these and perhaps other reasons, gold reserves have made almost no direct contribution to financing the oil consuming areas' structural deficit vis-à-vis OPEC, which means that they have not been used to restrain the growth of international debts. Some indirect contribution has come from their use as explicit or implicit collateral for external borrowing. Besides the Italian example mentioned above, Portugal and South Africa have used this technique explicitly, and there have doubtless been other, unpublicized instances.

3.3. A number of OPEC and of non-oil developing countries have nonetheless acquired or expanded official gold holdings, most notably Indonesia (see following paragraph). A more innovative move was the cautious step in the constitution of the European Monetary System towards partial re-mobilization of western European countries' gold reserves. For the time being, the remobilization directly impinges only upon the EMS members themselves, because ECUs — the reserve units created against the deposit of 25 percent of members' national reserves, gold being valued at a market-related price — cannot be acquired by outsiders. However, if creation of ECUs increases the reserves which EMS members feel are available for spending outside the System, then there is already a wider effect. On the other side, gold has been formally written out of the IMF Articles of Agreement; and while the IMF has now disposed of one-third of its former gold holdings — half of it by restitution to member countries and half by sale on the market, the "profits" over book value being applied to the new IMF Trust Fund — this still leaves two-thirds of those former holdings unavailable for use in IMF operations and contributing at best some last-resort commercial underlay to the Fund's balance sheet. Tentative proposals were made in 1980 to sell more IMF gold on the market, perhaps employing the proceeds to finance possible exchange losses by an IMF Substitution Account which would hold dollar assets and SDR liabilities. These proposals found little favour, and in any case the US authorities turned against their own scheme for a Substitution Account, which was accordingly dropped.

3.4. Thus the picture is one of miscellaneous experimentation rather than any systematic or widely favoured approach. The steam has, however, gone out of efforts to remove gold totally from the monetary system; and in 1980 accrual of monetary gold reserves from new production reached a scale not seen since the mid-1960's. In the words of the BIS Annual Report (June, 1981): "Excluding the US Treasury auctions and the IMF restitution transfers, countries' gold reserves went up by 211 tons last year, as against 26 tons in 1979. In other words, for the first time for many years there were sizeable additions to gold reserves from new production and market purchases. South Africa's gold reserve increased by 66 tons, owing partly to the unwinding of earlier swaps with commercial banks, while that of Canada was reduced by 37 tons, partly for the minting of gold coins. More significant, however, was the movement in the official gold stocks of developing countries (both OPEC and non-OPEC) which, excluding IMF restitutions, increased by 178 tons (in 1980), as against 22 tons in 1979. OPEC countries added 106 tons to their published gold reserves, with Indonesia alone accounting for 66 tons, while in non-OPEC Latin America and Asia there were increases of 41 and 28 tons respectively." (loc. cit., p. 140).

In 1980 gold did thus make a modest contribution to financing the OPEC payments surplus. The possibility of encouraging this trend should give added impetus to the search for methods of remobilizing gold reserves in the hands of central banks.

4. A Return to Pegged Rates?

As for the second question posed in section 3, my personal view is that a return to worldwide pegged exchange rates is desirable, but not in the foreseeable future attainable. It is desirable because floating rates have fluctuated excessively and have thereby, in conjunction with rapid inflation, undermined the efficacy of the price mechanism, to the serious detriment of employment levels and economic welfare generally. The exchange-rate developments of the 1970's have thus given renewed support to Ragnar Nurkse's classic study of the 1930's (*International Currency Experience,* Geneva, League of Nations, 1944) rather than to the criticisms of Nurkse advanced by Milton Friedman.[2]

An additional welfare loss arises from the diversion of scarce resources, especially skilled manpower, to speculative, hedging, trading and even manufacturing activities, whose main *raison d'être* is the exploitation of price fluctuations and expected inflation. Such activities, with their vested

[2] See Milton Friedman, "The Case for Flexible Exchange Rates" in *Essays in Positive Economics* (Chicago, 1953), p. 176. Friedman's methodological points, it should be said, retain a good deal of validity, even if Nurkse's judgement is supported by recent experience. In any case, those who wish to blame governments rather than markets for the instability of exchange rates now have new ammunition in the form of rational expectations.

interest in instability, include a good deal of gold trading, particularly in the gold futures market, which is nothing more than a gambling casino. I mention this point *inter alia* because it underlines the hollowness of the criticism formerly levelled at the monetary role of gold, that it involves a needless diversion of real resources to first producing gold and then locking it up in bank vaults.

4.1. Global pegged rates are, nevertheless, unattainable for the time being — in spite of the fact that, if official gold reserves are valued at the sort of market prices prevailing since 1978, their quantitative relationship to world foreign exchange reserves is broadly similar to what it was in the late 1960's; for instance, at the end of 1980 total national gold reserves were valued at about $550 billion, and other forms of reserves at $413 billion[3].

The impossibility of a return to pegged rates is due not to international mobility of liquid funds nor because of OPEC payments surpluses, but because of (i) rapid world inflation, which makes a pegged gold price impossible, and (ii) absence of the political will and other prerequisites for repegging exchange rates either to gold or to some man-made instrument such as SDRs.

4.2. I do not wish here to add anything further on the political aspects, but the point about inflation needs amplifying. The need for a higher gold price in the 1960's stemmed ultimately from the rise in nominal incomes during and after World War II; or in other words, from the moderate, not rapid, inflation that had characterized the previous thirty years. The gold price had, of course, been increased once before; the IMF Articles made provision for it to be adjusted, if necessary, in terms of all currencies simultaneously; and there is no reason to suppose that further increases at lengthy and irregular intervals (say, once in a generation) would have made the system unviable. World inflation at the rates experienced in the 1970's, however, is another matter. If inflation is such that no commodity price can be relied on to stay constant for a significant period of time, then currencies cannot be pegged to gold either. Attempts to do so will cause the gold market to be dominated by speculation on future price movements. The reserve accrual mechanism will prove hopelessly unreliable. So, by our earlier analysis (sections 2.7 and 3), major currencies will be left with no option but to float.

4.3. The foregoing argument, it should be noted, makes rapid worldwide inflation a sufficient condition for floating exchange rates, even if there are no substantial divergences among the national inflation rates of individual countries. But suppose now that such divergences exist. Might they be an

---

[3] See Appendix B.

alternative sufficient condition for floating? An affirmative answer here is often taken for granted, but in my view is probably incorrect, for the following reason. To affect the character of the system as a whole, and not merely the position of one or two countries, divergences in the rate of price increase must be widespread and varied. So they will be compatible with an absence of world inflation *on average* only if the price level in some countries is actually falling. In that case it is also likely that one major country or group of countries will have approximately stable prices, and therefore be able, if it wishes, to establish a par value in terms of gold. Once the system is anchored in this way, differences in rates of inflation/deflation can be accommodated by means of adjustable or crawling pegs, altered in accordance with agreed, Bretton Woods-type criteria, rather than by floating.

4.4. This line of reasoning also indicates why international mobility of capital is not a decisive obstacle to the maintenance of a pegged-rate system. If international flows of funds create pressures in particular cases, the management of exchange markets and of parity changes can be adapted accordingly. Devices such as crawling pegs or temporary floating by individual currencies are quite compatible with the maintenance of a gold-based system, provided only that gold convertibility is maintained by the centre country or core of countries.

4.5. It must be emphasized, however, that the cure of rapid inflation (and a "cure" implies that there is no long term cost in terms of employment losses) is a pre-requisite for the re-establishment of a gold-based system. Having once acquired the power to alter the supply of money independently, policy-makers cannot then renounce this power and hand it back, so to speak, to the gold producers. Once price stability is regained, re-establishment of a gold-based system can provide valuable policy guidelines for the maintenance of such stability in subsequent years; but who knows whether the political will to go back to gold will then be present?

5. Appendix A

Milton Gilbert on the Monetary Role of Gold since 1945

by Peter Oppenheimer and Michael Dealtry

(Extract from Editors' Introduction to Milton Gilbert, *Quest for World Monetary Order: The Gold-Dollar System and Its Aftermath,* a Twentieth Century Fund Study. Copyright © 1980 by the Twentieth Century Fund. Published by John Wiley & Sons Inc., New York)

Gilbert first formulated his ideas on the subject in a long essay written in the mid-1960's, which was subsequently published in two parts: the first as Princeton Essay in International Finance No. 70, *The Gold-Dollar System: Conditions of Equilibrium and the Price of Gold,* and the second as a Charles C. Moskowitz lecture, "The Discipline of the Balance of Payments and the Design of the International Monetary System" (in *Inflation: The Charles C. Moskowitz Lectures for 1970,* New York University Press). The present posthumous book owes much to that earlier essay, although it ranges far more widely.

Gilbert's contribution was to fill an important gap in the analysis begun by Robert Triffin in *Gold and the Dollar Crisis* (Yale University Press, 1960) of the relationship between the US balance of payments and the condition of the international monetary system as a whole. Triffin had noted that, by running an overall deficit in its balance of payments, the United States was acting as a net supplier of foreign exchange reserves to the rest of the world. The "Triffin dilemma" focused on the fact that this process was weakening the external liquidity position of the United States and could not continue indefinitely without calling into question the dollar's convertibility into gold at $35 an ounce. Hence, it was argued, the United States would sooner or later have to take steps to eliminate its external deficit; the world would thereupon face a shortage of international liquidity and accordingly would have to choose among a higher gold price, abandonment of fixed exchange rates in favor of floating ones, and an increasingly centralized system controlled by an International Monetary Fund (IMF) with greatly extended powers over the policies of its member countries.

This account, impressive though it was, had little to say about the causes of the US deficit or whether the causal mechanism was important for the system's future development. Was it deliberate philanthropy on the part of the United States, a kind of extension of Marshall Plan aid? Or fortunate accident? Or was the dollar basically overvalued in some long run sense? And if it was, in what sense? Gilbert provided the main outlines of an answer to these questions. The roots of the US deficit lay neither in philanthropy nor in coincidence but in the condition of the international monetary system itself, and specifically in the deficient inflow of new monetary gold at the prevailing official price coupled with the world's reliance on gold and dollars to provide for the expansion of international reserves. To satisfy the demand of the rest of the world for increments in reserves, the United States was not only bound, if it failed to run a payments deficit of its own accord, to be pushed into deficit by the policies of other countries; in addition, at the existing price of gold, this deficit was bound to involve a persistent drain on the US gold reserve because the US Treasury, by standing ready to sell gold to foreign monetary authorities against dollars, was acting in effect as buffer-stock manager for the underpriced gold reserves of the world as a whole. In this respect, the dollar's problems arose from its overvaluation — but overvaluation, in common with all other currencies, vis-à-vis gold, not overvaluation vis-à-vis other major currencies.

The picture was complicated by the fact that in the late 1960's, as a consequence of the war in Vietnam and other factors, the dollar also gradually became overvalued in relation to other currencies. The two causative elements in the US deficit were difficult to disentangle — especially, of course, by those who refused to recognize that gold had played any part in the first place. The IMF Articles had used the term "fundamental disequilibrium" to denote a situation in which a change in the exchange parity of an individual currency was appropriate. Gilbert's insight was to see that this term also could be applied by analogy to the monetary system as a whole, to denote a situation in which a change in the gold value of all currencies, as opposed to the parity of a single currency, was appropriate. Such a situation affected the United States in a unique way because of the dollar's central role in the system. But just as the IMF Articles provided for alteration of the par value of a single currency (including the par value of the dollar) if that currency was in fundamental disequilibrium, so they also provided for a "uniform change in par values", that is, for a general rise in the price of gold, to remedy what Gilbert called a fundamental disequilibrium of the system itself.

It followed from this analysis that Triffin was mistaken in his conception of a world not having to choose its future monetary constitution until the US external deficit had been corrected. The choice would be essentially preempted by the United States itself in deciding what to do, or not to do, about its deficit — that is, whether to devalue the dollar against gold or to abrogate its convertibility into gold (it was evident that Triffin's third option, centralized management of the world monetary system, was not a practical possibility since it would require a large step in the direction of world government).

Gilbert had much more than an academic's concern with the balance-of-payments adjustment mechanism for the United States. Throughout his years spent abroad as an international official, he never ceased to be a devoted American. And he perceived that an adamant refusal to raise the dollar price of gold would not merely render the fixed-rate system unworkable, but also would undermine the position of the US dollar in particular, and was therefore contrary to the national interests of the United States. He received no thanks from US officialdom for saying so, as he himself recalled at his farewell dinner at the BIS in 1975. "When I look back now over fifteen years, I have to admit that in the main thing I tried to accomplish I failed. And that was to convince my own country to take the action necessary to maintain the status of the dollar, to correct the balance-of-payments deficit and to abide by the Bretton Woods system... It's not only that they did not want to follow me; they were so caught up by the political difficulty in the United States, they couldn't clear their minds enough to even understand me... You know, they came to treat me not only as wrong, but as a kind of a half-traitor, lined up with Rueff and de Gaulle. Why? Just because I argued that we didn't have to sit on our hands and lose billions of dollars every year — to a total of over sixty-five billions, about five Marshall Plans."

Gilbert also was incensed by the acquiescense, or worse, in this process on the part of the great majority of his fellow American economists, an attitude that he felt reflected both intellectual and political shortcomings. Gold had long ceased to be a part of the domestic money supply, and the dollar was functioning as the principal intervention and reserve currency of the international system. In these circumstances, the idea that the character of the system still somehow hinged on gold was intellectually repugnant to most economists, who therefore did not take the trouble to try to understand it. Of course, the dollar's attraction as a currency was connected with its purchasing power in terms of US goods and services, and also with the facilities offered by the New York money market for the investment of liquid funds. Gold, however, enabled the dollar to stand at the center of a fixed-rate system because, as an inherently scarce, real commodity alternative to the dollar, it provided the United States both with a criterion of balance-of-payments discipline (namely, the dollar's convertibility into gold), and with a vital instrument to help observe that discipline and keep the dollar strong (namely a change in the dollar's par value).

Some commentators, both anti- and pro-gold, appeared to think that the discipline of gold was an all-or-nothing matter: either gold was the sole international reserve medium of any consequence and it had a price in terms of major currencies that never changed or it could play no monetary role whatever. This view may have been a tribute to folk memories of the pre-1914 gold standard, but it showed little grasp of the foundations of the Bretton Woods system. A substantial change in the dollar price of gold was entirely consistent with — and in the end necessary for — maintaining the Bretton Woods system, provided only that such changes were not so frequent or so large that the gold market became continuously dominated by speculation about the next change.

With the abrogation of the dollar's convertibility into gold in 1971, the United States ceased to have a clear criterion of balance-of-payments discipline, a change reflected in the new presentation of US balance-of-payments statistics after 1976. Moreover, the US authorities lost most of the power that the gold link had given them to manage the US exchange rate and hence to maintain the strength of the dollar in the monetary system. Many academic economists welcomed the resulting move to floating exchange rates — and later added (as if it were self-evident) that the monetary consequences of higher oil prices after 1973 would have made floating inevitable anyhow. But there were also economists who argued that floating rates (and currency blocs) could have been avoided either by devising new options to the dollar in place of gold or by going over to a "dollar standard," in which all countries pegged on the dollar and the United States as the residual ("*n*th") country in the system would adopt a purely passive attitude to its exchange rate and balance of payments. This dollar standard was an armchair fantasy, particularly given the fact that a major spur to US action on the balance of payments in 1971 was the urgent wish of the authorities in Washington to force a correction of the dollar's overvaluation vis-à-vis other currencies. A return to fixed rates on the basis of IMF Special Drawing Rights (SDRs) — with dollar and other currency reserves restricted in amount and possibly phased out altogether — is something vaguer than an armchair fantasy because its mode of operation cannot be clearly visualized. In the gold-dollar system, there was a strong and direct link between the exchange-rate policy and the international liquidity position of the United States; a reduction in the dollar's par value could always be made large enough to produce a decisive impact on US reserves. To have any foreseeable prospect of viability, a fixed-rate system based on SDRs must reproduce this feature; and the fact that SDRs are a fiduciary asset and not a commodity means that any blueprint to date has been far from doing so.

Exactly how much a timely devaluation of the dollar in the 1960's would have contributed to world monetary stability in the 1970's and 1980's is open to debate. But the prolonged disequilibrium of the fixed-rate system, culminating in a massive flight from the dollar and a mushrooming of central bank currency holdings, bears a considerable measure of responsibility for the inflationary outburst of the early 1970's — and, incidentally, for the extent of the subsequent increase in the market price of gold. In this connection, Gilbert's analysis also reconciles what were (and perhaps still are) often regarded as contradictory features of the international monetary scene: how could there be at one and the same time a shortage of international liquidity apparently requiring governments to search for new ways of creating reserves and an abundance of international liquidity permitting double-digit inflation by the early 1970's? The answer is that the shortage was never of total world liquidity but only of gold, and the gold shortage itself helped to produce a flow of dollars that ultimately became excessive not only from the standpoint of the US balance of payments but also from that of global reserves and global price stability. By the same token, a deliberate and timely increase in the gold price, adequate to correct the disequilibrium in the gold-dollar system, would not have been inflationary.

Persistent world wide inflation is the main economic reason why a return to a gold-based monetary system is ruled out for the foreseeable future. In the presence of two-digit inflation, gold cannot — any more than other commodities — be given a fixed price in terms of currencies. Attempts to restore gold's earlier monetary function would therefore have to be preceded by a return to price stability, not the other way around. Gold can help to maintain stability only when stability has been achieved. But questions of price stability apart, attempts to restore gold are ruled out by the political hostility of the United States and many other countries, particularly in the developing world. This hostility is reflected in the 1976 amendments to the IMF Articles of Agreement, which effectively abolished any role for gold in the working of the Fund.

Gilbert believed that exchange reserves are likely in the future to be a mixed bag. The question for the 1980's — just as for the 1960's — is not the volume but the composition of international reserves. Claims denominated in SDRs, and even SDRs themselves, may play a more prominent role than before, along with currencies other than the dollar, as well as dollars and gold. But all this will still be in the context of a floating-rate system among the major currencies. Moreover, under such a system stability is likely to prove an elusive goal for several reasons, not least because of the absence of satisfactory criteria for balance-of-payments policy. Under floating rates, "fundamental disequilibrium" has no clear meaning; the line between autonomous and compensating capital flows becomes even more blurred than before; and a significant element of unpredictability in exchange-rate movements, both short and medium term, is a fact of economic life. Yet a worldwide return to fixed rates among market economies is now difficult to envisage.

6. Appendix B

Reserves, Sources and Uses of Gold

Table 1 - *Estimated sources and uses of gold*

Items	*1977*	*1978*	*1979*	*1980*
	in metric tons			
Production	965	970	955	940
Estimated sales by Communist countries	450	450	290	90
Change in Western official gold stocks (— = Increase)	275	245[2]	600	—110
of which: Countries' gold reserves[1]	*— 70*	*—220*[2]	210	—250
Total (= estimated non-monetary absorption)	1,690	1,665	1,845	920

[1] Including gold restituted by the International Monetary Fund to its member countries, totalling 371 tons in 1977, 189 tons in 1978, 180 tons in 1979 and 37 tons in 1980, as well as gold swaps by members of the European Monetary System against ECUs during 1979 and 1980.

[2] Including 65 tons of gold transferred from the Japanese Ministry of Finance to the Bank of Japan which had not previously been counted as part of world gold reserves.

Source: Bank for International Settlements, *Annual Report 1981*.

Table 2 - *Changes in global reserves, 1978-80*

Areas and periods	Gold		Foreign exchange	IMF reserve positions	SDRs	ECUs	Non-gold total
	in millions of ounces	*in billions of US dollars**					
Group of Ten countries and Switzerland							
1978	3.2	51.6	35.6	— 2.6	— 0.1		32.9
1979	— 95.0	199.2	— 31.3	— 2.2	3.8	42.2	12.5
1980	— 1.1	47.9	9.7	3.1	— 1.0	20.9	32.7
Amounts outstanding at end 1980	*739.5*	*436.0*	*118.5*	*12.1*	*10.2*	*63.1*	*203.9*
Other developed countries							
1978	— 1.0	5.5	10.1	0.3	0.3		10.7
1979	0.6	28.3	1.1	— 0.1	0.2	0.7	1.9
1980	2.5	7.6	3.4	0.6	—	— 0.1	3.9
Amounts outstanding at end 1980	*96.7*	*57.0*	*37.0*	*1.9*	*1.2*	*0.6*	*40.7*
Developing countries other than oil-exporting countries							
1978	2.9	3.9	13.7	0.4	0.3		14.4
1979	2.2	17.7	8.1	0.3	1.1		9.5
1980	2.8	5.5	— 1.2	1.0	— 0.5		— 0.7
Amounts outstanding at end 1980	*60.8*	*35.8*	*67.3*	*2.2*	*2.1*		*71.6*
Total oil-importing countries							
1978	5.1	61.0	59.4	— 1.9	0.5		58.0
1979	— 92.2	245.2	— 22.1	— 2.0	5.1	42.9	23.9
1980	4.2	61.0	11.9	4.7	— 1.5	20.8	35.9
Amounts outstanding at end 1980	*897.0*	*528.8*	*222.8*	*16.2*	*13.5*	*63.7*	*316.2*
Oil-exporting countries							
1978	1.9	2.5	— 14.5	— 0.8	0.2		— 15.1
1979	0.3	11.2	15.4	— 1.8	0.7		14.3
1980	3.4	4.4	19.8	1.3	0.2		21.3
Amounts outstanding at end 1980	*40.6*	*23.9*	*90.3*	*5.3*	*1.6*		*97.2*
All countries							
1978	7.0	63.5	44.9	— 2.7	0.7		42.9
1979	— 91.9	256.4	— 6.7	— 3.8	5.8	42.9	38.2
1980	7.6	65.4	31.7	6.0	— 1.3	20.8	57.2
Amounts outstanding at end 1980	*937.6*	*552.7*	*313.1*	*21.5*	*15.1*	*63.7*	*413.4*

* Gold reserves valued at market prices.

Source: Bank for International Settlements, *Annual Report 1981*.

CHAPTER 19

THE EUROPEAN MONETARY SYSTEM AND GOLD*

FREDERIC BOYER DE LA GIRODAY**

1. Introduction

The weight of history on the shaping of policies of nation states is one of the most striking features of international relations. Considerable inertia derives from it even though in all fields — except in those of politics and institutions — underlying factors are changing so widely, deeply and rapidly that our age has been labelled as a "revolutionary century"[1].

In the more limited field of contemporary monetary arrangements, which spread over a relatively short time on the scale of history, the weight of precedent is as great. As in biology, changes proceed very rarely by way of radical innovation and depend rather on the slow maturation of ideas; elements, once useful, outlive their initial purpose.[2]

In his famous book *Gold and the Dollar Crisis*, Robert Triffin presented a modernized version of the proposal of Keynes for an international clearing union. Many ideas including the pooling of gold found their way into official blueprints. A central idea, the possibility for an international organization to create fully fledged money, was never implemented, except in a very limited manner, with the creation of the SDR, which so far has not played a

---

* The views expressed in this paper are those of the author and should not be attributed to the EEC Commission. The place assigned to this presentation by the organizers of this Conference is in the section called "National and Supranational Monetary Authorities". As far as the EMS is concerned, neither of the two attributes are appropriate.

** Mr. Frédéric Boyer de la Giroday died on December 29, 1981. His paper was presented at the World Conference on Gold by Mr. Thomas Webb.

[1] On this see the monumental history of Fernand Braudel, *Civilisation matérielle, Economie et Capitalisme*, Vol. 3 p. 543.

[2] On the weight of history on the present complexities of the Articles of Agreement of the International Monetary Fund see J.J. Polak's, *Thoughts on an International Monetary Fund Based Fully on the SDR*, IMF 1979, p. 4. Polak states explicitly that "the explanation is historical in nature".

significant role in international monetary affairs. In a very lucid pamphlet, J.J. Polak recently (1979) developed certain of these ideas in a still more modernized form ("Thoughts on an International Monetary Fund based fully in the SDR"). Authorities have repeatedly failed to act upon them.

The European Monetary System (EMS) in its present transitional phase, and recent controversies about gold on which attention will be focussed in this presentation, provide us with good examples of conservatism in international monetary affairs.

However questionable the circumstances in which they started in 1971, moves have been made to phase gold out of the international monetary system, but although it no longer holds a central or pseudo-central place in international relations as it once held in the Bretton Woods system, it is still a part of the reformed IMF's accounts and transactions[3] and the initial actions designed to reduce that part through the sale of gold have been stopped, although the Fund is repeatedly in need of liquid funds. The same is true of the United States' moves in this direction. Moreover, with few exceptions, central banks are still clinging to their gold even though some of the countries concerned are heavily indebted. Finally, a fairly powerful thrust (judging from the intellectual[4] and political[5] standing of its proponents) has emerged lately, as illustrated by moves tending to give back to gold all or part of its past monetary lustre.

Gold is a part, albeit an ancillary part, of the European Monetary System. This will be examined in section 2. At this stage it is not known whether and what role gold will be called upon to play in the promised, but delayed, "definitive" stage of the EMS. It is likely that some role will be assigned to it and section 3 will examine the theoretical options offered to the negotiators. In the same section, an idea will be presented which has not been taken up lately: the introduction of gold in the composition of the ECU. This idea is far from new since it was a subject of debate in the late sixties and early seventies at the time of the long search for some substitute for the gold-dollar system. At that time the insertion of a Composite Reserve Unit — the CRU — in the system was proposed but never taken up at official level. It will be found that the introduction of gold in the ECU is not a practical proposition.

---

[3] See Articles IV and V of the Articles of Agreement of the IMF 1978 edition.

[4] See *The 1981 Per Jacobsson Lecture* by J. Zijlstra, President of the Nederlandsche Bank, pp. 8 to 10, typed version dated July 30, 1981; less recent writings include those of Jacques Rueff and Fred Hirsch, *The Role and the Rule of Gold: An Argument*, Essay n° 77, December 1969, Princeton Essays in International Finance; and Milton Gilbert, *The Quest for World Monetary Order*, posthumously edited by Peter Oppenheimer and Michael Dealtry, John Wiley & Sons Inc., New York, 1980.

[5] See Resolution of the Congress of the United States appointing a Gold Commission, 1981.

Certain conclusions will be presented in section 4 which will relate present issues to the natural conservatism of authorities and ordinary people in money matters and to the boundless human capacity to entertain illusions about solving difficult psychological and political problems — in this case chronic inflation — with the help of some "deus ex machina", in this case a metal whose monetary virtues belong to another — long past — age.

2. Gold and the EMS in Its Present Stage

2.1. *The Need for Stable but Adjustable Rates in the EEC*

Attempts at organizing a monetary system within the Community have been first and foremost of a political nature. It was not explicitly so originally (1958) because both the world and Europe had sufficiently well functioning monetary systems (the International Monetary Fund and the European Payments Union) which could meet economic and monetary needs of the Community, but that the EEC required its own organization in the field of money could be detected from its broad objectives (the creation of a Common Market, in which goods, services, labour and capital would move freely). This requirement became still more obvious when the Bretton Woods order broke down in the early seventies.

When a fluctuating rate regime was forced upon the world in the late sixties and early seventies (dollar instability, immobilism in the Bretton Woods System and the first oil crisis), floating had had many precedents in the checkered history of international monetary relations (in the last instance in the 1930's). Its intellectual revival was started in the early 1950's by Milton Friedman and was extended in the 1960's and 1970's to a very important sector — perhaps a majority — of the academic world. As often in the history of economic ideas, events — in this case pronounced inflationary developments — gave a powerful impetus to the force of inductive arguments. Such arguments have been somewhat battered by developments over the last few years but they have retained a great deal of force, at least when expounded from a general (world) point of view.

As time and experience were to show, after that regime had invaded intra-European monetary relations, fluctuating rates appeared more and more inappropriate for a region like the EEC whose hallmark — beyond institutional arrangements — is a very high degree of de facto integration by way of current and probably capital transaction (although integrational via capital movements is not amenable to precise statistical measurement in present circumstances). A few figures will illustrate the importance of trade for the economies of these countries and the share of intra and extra trade.

Exports (or imports) represent between 20 percent and 25 percent of GNP for the larger economies of the EEC and 50 percent of the GNP of the Benelux group; of these figures at least half of the trade of these countries goes to their EEC partners. A policy tending to maintain the exchange rates of these countries in their reduced margins of fluctuation is therefore of crucial importance; but the realization of this objective must be weighted against the limits of the socio-economic constraints that the nations concerned are prepared to observe. That "only" two general realignments of exchange rates took place since the establishment of the system in March 1979 — one in September 1979 and the other in October 1981 — has been considered as a success. This has been due partly to convergent monetary policies and partly to a favourable constellation of circumstance (a weak Deutsch Mark and a strong French Franc during most of the period between these two dates) of a fortuitous character.

The functioning of the EMS has shown that institutions are an important factor in monetary management but that the dictates of institutions cannot be pushed beyond certain limits set by socio-political acceptability. The whole post-war history of international monetary relations, of which the major milestones are summarized in the next section, illustrates the very relative success of maintaining "stable but adjustable" rates of exchange between currencies. This should be an important consideration in the present debate about the restoration of a gold standard.

2.2. Supporting Mechanism and the Role of Gold in Earlier Monetary Systems

Basically all the trade and exchange systems devised or operated since the war were based, explicitly or implicitly, on the notion of "stable but adjustable" rates. All of these have had a supporting mechanism designed simultaneously to improve stability of rate (via intervention) and to limit inflationary propensities. The main features of the system which preceded the EMS are set forth hereafter.[6]

A) The Keynes Plan for an International Clearing Union was never implemented but contained seminal ideas which influenced all the later constructions. Its main characteristics were as follows:

(i) No pooling of gold or currencies accumulated in the past.

(ii) Over and above traditional means of settlement (gold and currencies) in their possession, members were provided with quotas (or "overdraft facilities") expressed in bancors (whose value was to be fixed by the

---

[6] Most of the information provided in this section is derived from Robert Triffin's *Europe and the Money Muddle*, Yale, 1957.

Governing Board), enabling them to settle deficits via transfers to creditors. Quotas were to be set at 3/4 of countries' average trade turnover (exports plus imports).

(iii) Borrowing in excess of quotas was possible at Governing Board discretion (weighted voting system) against the surrender of gold by the debtor country or other (policy) conditions. There were no acceptance limits to creditors' obligations to receive bancors, and no time limit on outstanding creditors' positions, but borrowing beyond half of quotas was to be subject to "consultation" with the Governing Board (no sanctions envisaged at that stage).

(iv) Thus the role of gold was limited, but no provisions for phasing it out were set forth, although the whole scheme was biased against gold (Article 26 of the scheme said: "What, in the long run, the world may decide to do with gold is another matter"). For the first time in history an international organization would be empowered to create international money (the bancor). The Board could reduce or increase quotas uniformly (i.e. control liquidity of the system).

B) The original IMF system took effect, in practice, in 1958 after return to convertibility, i.e. several years after inauguration on December 27, 1945:

(i) It was (and still is) based on a quota system: quotas were payable in gold (25 percent) and in members' own currency (non-interest-bearing liabilities, i.e. IOU's).

(ii) The amount of each quota determined members' borrowing facilities and lending obligations; time limits were set for outstanding debtors' positions. Borrowing beyond the "gold tranche" was subject to conditionality.

(iii) The role of gold was important in that gold holdings were partially pooled and could be sold for currencies of which there would be a shortage. Provisions were made for general change in total amounts of quotas, in principle every five years, and for changing the price of gold in terms of all currencies. This latter provision was never implemented and certain critics of the IMF organization place a great deal of responsibility of this fact for the collapse of the Bretton Woods system.

(iv) The role of gold in the present (as from January 1978) functions of the IMF has been largely reduced and to a large extent replaced by the Special Drawing Right (see Articles IV and V of the IMF).

C) The EPU system:

(i) No pooling of reserves accumulated in the past.

(ii) Based on a system of quotas (equal, for each member, to 15 percent of total visible and invisible trade) measuring borrowing rights and financing obligations within the limits mentioned below.

(iii) Minor different provisions notwithstanding, debtors could settle their

debts by a graduated system of combined gold (or dollar) payments and by drawing on credit from the surplus countries. The proportion of gold payments rose as cumulative deficits increased; overall lines of credit were fixed (unless otherwise determined by the OECD Council) to 40 percent of each country's quota. Reverse, but asymmetrical provisions, were stipulated for creditors' financing obligations. The asymmetry between debtors' facilities and creditors' obligations was catered for by the possibility of drawing from a 350 million dollar capital fund contributed by the United States in the framework of the Marshall Plan. This contribution would serve, also, as a guarantee to the creditors against possible defaults.
(iv) The EPU was terminated at the end of the year 1958 and replaced by a rather insignificant system called the European Monetary Agreement. The IMF provisions were then implemented in accordance with its Article VIII on convertibility.

2.3. *The EMS*

A) Antecedents: the first impulsions at creating a Community Monetary Organization were given in December 1969 in the usually rather vague but occasionally grandiloquent language of the communiqués issued after Community Summits (later on called European Councils) at The Hague Conference of Chiefs of State and Governments, when an agreement "in principle" was reached regarding the creation of an economic and monetary union to be achieved by 1980. This started a long drawn-out process of negotiations which were complicated by the upheavals in the international monetary system. The initial landmark in these negotiations was the setting up of the system set up by the Basle Agreement of 24 April 1972 for the narrowing of the margins of fluctuations between the Community currencies: the snake (margins of 2.25 percent) in the (dollar) "tunnel" as defined by the Smithsonian Agreement of December 1971 (margins of 2.25 percent). This system had a rather checkered life. The dollar tunnel disappeared with the emergence of generalized floating in 1973; in the meantime the Pound Sterling, the Irish Pound and the Danish Krone had joined the snake in 1972, before their countries had formally acceded to the Community; and then two of these currencies had left it; so, later, did the Italian Lira, while the French Franc left the system, came back subsequently and then left it again.

The hard core of the system was made up of the German Mark, the Belgian Franc and the Luxembourg Franc, the Dutch Guilder and, for a while, the Norwegian Krone and the Swedish Krone.

The snake system was relatively simple; beyond the exchange rate arrangements mentioned above, it contained:

— a very short term financing mechanism (VSTF);

— a short term (three months renewable) monetary support system (STMS), administered initially by the Committee of Governors of central banks of the EEC countries, and subsequently by the European Monetary Cooperation Fund (composed of a governing body composed of Governors of central banks); the granting of credit was largely automatic;
— a medium term financial assistance system subject to decisions of the Council of Ministers of the Community.

The accounting of the system was held by the Fund in terms of the European Currency Unit (ECU) created earlier. The ECU is a basket type unit of account, serving also as "settlement instrument and reserve vehicle for the purposes of the system". There were no explicit provisions on gold, except that in the early stages the unit of account was expressed in terms of gold, later, and for a while, transmuted into an ECU grid of parities. A European Monetary Cooperation Fund (EMCF) was created with a view to catering for the accounting of the system.

B) The mechanics of the EMS and the role of gold

(i) Prologue: The origins of EMS. The establishment of the EMS followed a period of hectic negotiations (January-October 1978) which took place after a decision of principle had been taken at the Summit meeting held on July 6-7, 1978 in Bremen. Attendance at such meetings, usually preceded by secret encounters, is limited to persons holding high office and the precise background of official communiqués and/or resolutions is often lacking.

The crucial passages of the Bremen Communiqué on monetary organization is reproduced hereafter:

"... 2. Monetary policy

Following the discussion at Copenhagen on 7 April the European Council has discussed the attached scheme for the creation of a closer monetary cooperation (European Monetary System) leading to a zone of monetary stability in Europe, which has been introduced by members of the European Council. The European Council regards such a zone as a highly desirable objective. The European Council envisages a durable and effective scheme. It agreed to instruct the Finance Ministers... to formulate the necessary guidelines for the competent Community bodies to elaborate by 31 October the provisions necessary for the functioning of such a scheme — if necessary by amendment."

An annex to the Communiqué provided that:

"... 2. An initial supply of ECUs (for use among Community central banks) will be created against deposit of US dollars and gold on the one hand (e.g. 20 percent of the stock currently held by member central banks) and member currencies on the other hand in an amount of a comparable order of magnitude.

The use of ECUs created against member currencies will be subject to conditions varying with the amount and the maturity; due account will be given to the need for substantial short term facilities (up to 1 year)."

The Bremen project outlined above was implemented by the European Council of December 5, 1978 on the basis of reports by the Monetary Committee, the Committee of Central Bank Governors and the Council which produced interpretations of the rather sibylline provisions quoted above, parts of which (e.g. the deposits in member currencies) were left in abeyance. Its main provisions regarding the mechanism supporting the exchange system are summarized below. Exchange arrangements will not be dealt with in detail. They are rather complex in that they consist of a double system: a straight bilateral-nominal rate system which is the same as that of the snake arrangement and a system incorporating a "divergence indicator". The latter system resulted from a controversy about the sharing of the burden of adjustment between debtors and creditors. A compromise was reached which purported to assure a balanced distribution of the burden of adjustment through the "divergence indicator", but this had certain features which, in practice, reduced its significance.[7]

(ii) The ECU creation system. ECU creation has allowed a mobilization of gold. The system is based on three month renewable swaps, which together with the pricing convention has resulted in a large growth in the number of ECUs since the start of the EMS. Each quarter the central banks swap 20 percent of their gold and dollar reserves with the EMCF and receive ECUs in return. For the dollars the quantity is given by the most recent market exchange relationship between the ECU and the dollar calculated through the basket definition; for gold the average of market prices over the preceding six months is compared with the most recent market price and the lower of the two is used. Table 1 shows the global results of these arrangements.

It is readily apparent that the dramatic growth in the number of ECUs, from 23.3 billion to nearly 50 billion at the maximum in the April-July 1981 quarter, was caused primarily by the rise in the market price of gold. It was

---

[7] The divergence indicator provides data that continuously supply a picture of the relative positions of the currencies. Normally it can be expected that a divergence threshold, which is set at 75 percent of each currency's maximum deviation from the other currencies, would be reached before there are severe strains within the system. The indicator is the result of a compromise on the symmetry of adjustment question and is expected to provide a simple early warning device. When a divergence threshold is crossed, there is a presumption that the authorities of the country concerned will correct the situation by "adequate measures" and/or that an examination of the underlying conditions would take place within the appropriate bodies. A complete description of the European Monetary System has been published in European Economy, Commission of the European Communities, Offprint No. 3, July 1979. Further analyses of the functioning of EMS have been presented in the *Annual Report 1980* and in the *Annual Report 1981* of the Directorate General of Economic and Financial Affairs.

Table 1 - *ECU creation through the swap system*

Swap period beginning	*Gold deposits (million ounces)*	*Dollar amount (billions)*	*Gold rate (ECU per ounce)*	*Dollar rate $ = 1 ECU*	*ECU equivalent (billions)*		
					Gold	*Dollars*	*Total*
April 1979	80.7	13.4	165	0.75	13.3	10.0	23.3
July 1979*	85.3	15.9	185	0.73	15.8	11.6	27.4
Oct. 1979	85.3	16.0	211	0.70	18.0	11.3	29.3
Jan. 1980	85.5	15.5	259	0.69	22.2	10.7	32.9
April 1980	85.6	14.4	370	0.77	31.7	11.1	42.8
July 1980	85.6	13.7	419	0.70	35.9	9.6	45.5
Oct. 1980	85.6	13.9	425	0.71	36.4	9.9	46.3
Jan. 1981	85.6	14.5	447	0.75	38.3	10.9	49.2
April 1981	85.7	14.2	440	0.84	37.7	12.0	49.7
July 1981	85.7	12.7	406	0.97	34.8	12.3	47.1
Oct. 1981	85.7	11.5	402	0.91	34.5	10.5	45.0

* The Bank of England deposited 20 per cent of its gold and dollars as from July 1979.

Source: *Monthly Reports of the Agent for the European Monetary Cooperation Fund*

fortunate that the period used for calculating the average price was longer than the extreme peak in the gold market or the price would have risen much higher than 447 ECUs per ounce and the following drop would have been larger. Up until now the effects of the dollar price and volume movement have been swamped by these gold price variations.

The distribution of ECUs between participants like the total quantity of ECUs is dependent on these outside factors. The number of ECUs that each receives and the change in that quantity between any two swap operations is determined by the proportions of gold and dollars that they happen to hold in their reserves. The following purely hypothetical and rather extreme case

Table 2 - *Percentage change in ECU holdings between July 1979 and July 1981*

Central Bank of	*Change %*
Belgium	+ 105.5
Denmark	− 11.3
Germany	+ 54.8
France	+ 86.5
Ireland	− 6.2
Italy	+ 66.8
Netherlands	+ 87.5
United Kingdom	+ 20.2

N.B.: The July situation has been retained because that was the time when the UK joined the deposit scheme while staying out, until now, from the exchange arrangements.

provides an example. Two central banks, one of which swapped 0.5 million ounces of gold and 2.6 billion dollars and the other of which swapped 10.3 million ounces of gold and 140 million dollars, would both have received in return the same quantity — 2 billion — of ECUs in the swap operation beginning in July 1979. But two years later, the first would have received 2.7 billion and the second 4.3 billion ECUs from the transfer of exactly the same amount of gold and dollars.

Table 2 shows the actual change in the amounts of ECUs received by each central bank against the swap of gold and dollars.

As a reserve asset and settlement instrument, the role of the ECU within the existing system has in fact been limited. Its use has been confined to a small part of the settlements, because participants would as well use other instruments which are not in the sphere of influence of the system, and its supply is determined by a once for all rule and is consequently not controllable for the purposes of the system. Use is limited because of lack of control. Potential creditors are unwilling to accumulate ECUs and reluctant to give the ECU a larger role. To gain control over the system it would be necessary not only to increase the use of the ECU but to move towards the position where all Community official settlements are made in ECUs.

(iii) Credit arrangements. At the start of the existing phase the quantity of credit was increased substantially, but the main features of the credit arrangements already existing in the Community were unchanged.

Apart from the very short term financing, which as its name implies should normally be classified as financing rather than credit, these arrangements have remained unused, and the system has thereby lost a potentially important instrument. If decisions about exchange rate changes are made (by the system as a whole rather than participants acting independently) taking into account other forms of adjustment, it would be consistent that financing decisions are made in the same way. Increasing the existing credit facilities was not a step in this direction to the extent that members' access to them is automatic, but in the event it has so far been unimportant as the short and medium term facilities remain unused.

Comparable figures are not available for the Netherlands, but for the other participants Table 3 shows that net official foreign debt as a percentage of GDP has increased over the period 1978 to 1980 for all except Italy.

Over the same period, the reverse positions of the participating central banks remained virtually unchanged. If these conditions were to continue to prevail, the Community facilities would continue to remain unused unless something similar to the idea of ECU exclusivity were adopted.

The short term monetary support is ironically, in view of the use that has been made of other credit, meant to be a system of first resort. It has only been used twice since 1970 and never since the start of the EMS. In both

Table 3 - *Net official debt as a percentage of GDP*

	1978	*1979*	*1980*
Belgium	— 1.2	— 3.5	— 9.2
Denmark	—12.8	—14.0	—16.6
West Germany	1.2	1.0	— 0.4
France	— 4.1	— 3.9	— 4.4
Ireland	—16.7	—21.3	—25.2
Italy	— 3.1	— 2.2	— 2.4

cases it was Italy that was granted a credit and in the second case it did not mobilize the facility. Despite being based on debtor and creditor quotas, there is the potential for a substantial degree of flexibility over the amount of credit which could be granted. The quotas determine the basic amounts available as credit. For example the central bank of one of the biggest countries may have up to 1.7 billion ECUs, but then up to another 4.4 billion is potentially available if circumstances warrant it. The duration is initially for three months, but this may be extended for a further two-three month period. The necessary decisions on these questions would be made by the Central Bank Governors.

The MFTA was set up in March 1971 specifically to give assistance to Member States with balance of payments difficulties. It can be granted for a period of up to five years and to an amount of 14.1 billion ECUs. Within its framework there are provisions for conditionality and for the financing to be made in tranches contingent on the satisfactory attainment of the conditions. It has therefore potential as a useful policy instrument, but it remains unused, and so there is little to go on in the way of case studies on how the conditionality could work. The necessary decisions are made by the Council of Ministers, in which according to practical Community decision-making procedures, the country applying for the credit, like the others, ultimately has a veto. The Ministers' decision is taken on the basis of a report from the Monetary Committee.

The very short term financing (in contrast to other mechanisms) was used especially in the early months of 1981. It is primarily a complement to the intervention arrangements. Any participating central bank which has had to intervene using another participant's currency (because the two were at their bilateral exchange rate limit) may not only obtain that currency but also need not settle until the end of a period of 45 days from the end of the month in which the intervention was made. These arrangements are automatic, but by agreement the financing period may be extended for two further 90 day periods. A considerable proportion of inter-Community intervention takes place before currencies reach their bilateral margins, and in these cases the VSTF arrangements may be used if there is agreement between the two central banks concerned. In all cases the possibilities for settlement are that the debtor central bank may settle at any time using the original intervention currency if market conditions reserve and allow it to buy back that currency with its own. Alternatively, if the financing period reaches maturity, it may settle using any acceptable reserve asset including ECUs. The VSTF therefore largely relies on bilateral decisions between participating central banks where it is not automatic, but it is incorporated into the system when the financing period is extended and through the use of ECUs.

(iv) ECU: supply and use. Participants have a right to use ECUs in conjunction with the VSTF subject to an acceptance limit, but ECUs may also be used by consent for any settlement between participants. Use has however been limited. Table 4 shows that a number of participants have not used any of their ECUs and that with the exception of Belgium the others have used only a small proportion of their holdings.

This limited use does not reflect the lack of disequilibrium between participants as it should if change in ECU holdings were to become a potential instrument of the system. As mentioned above, it reflects the fact that other relatively freely available instruments may be used in the system together with ECUs. In these circumstances further use of ECUs would not necessarily be in the overall interests of the system.

2.4. *Concluding Remarks on the EMS in Its Present Stage*

The above survey of the mechanics of the EMS amply demostrates the substantial imperfections of a system which was hastily negotiated on the basis of not very clearly defined terms of reference. The motivations were partly political and partly economic.

Once more an opportunity was lost in that the logical construction that could have been built on the basis of the ideas put forward by Keynes in the 1940's, by Triffin in the 1960's, and by Polak more recently, was not devised. The obsession that international organizations must not be empowered to create money[8] has played a crucial role in that outcome. That that obsession should still play such a role is extraordinary given the size which monetary authorities have allowed the Euro-currency system to assume.

Another oddity in the present state of affairs should be pointed out. In contradistinction to what happened thirty-five years ago when much more money than Keynes had provided for in his plan was wisely poured by the United States into the construction of Europe in the framework of the

[8] The creation of Special Drawing Rights in 1970 is hardly an exception. More than ten years after this innovation, the monetary utilization of the SDR is still extremely modest (see J.J. Polak, op. cit.).

Table 4 - *Net use (—) or accrual (+) of ECU as a percentage of ECU received through the swap operations*

End of month		*Belgium*	*Denmark*	*Germany*	*France*	*Ireland*	*Italy*	*Netherlands*	*UK*	*Total*
1979	April	— 8.2	+27.4	—	—	—	—	—	—	±0.6
	May	—14.7	+33.2	+0.8	+ 0.2	—	—	—	—	+1.1
	June	—15.0	— 1.8	+2.8	+ 0.2	—	—	—	—	+1.2
	July	—13.9	— 1.9	+2.8	+ 0.2	—	—	—	—	±1.0
	August	—14.0	— 1.9	+2.9	+ 0.2	—	—	—	—	±1.0
	September	—14.2	— 1.9	+2.9	+ 0.2	—	—	—	—	±1.0
	October	—12.9	— 2.3	+2.7	+ 0.2	—	—	—	—	±1.0
	November	—13.5	—26.9	+3.8	+ 0.2	—	—	—	—	±1.4
	December	—25.2	— 5.6	+4.2	+ 2.5	—	—	—	—	+1.9
1980	January	—27.9	— 9.8	+3.8	+ 4.8	—	—	—	—	±2.3
	February	—39.7	—21.1	+3.9	+ 9.6	—	—	+0.8	—	±3.3
	March	—54.1	—29.1	+0.8	+15.4	—	+3.5	+6.2	—	±4.5
	April	—43.3	—45.7	—1.4	+17.3	—	+2.6	+1.6	—	±4.3
	May	—37.6	—46.1	—1.5	+17.4	—	+0.9	—	—	±3.8
	June	—31.2	—46.5	—1.5	+15.8	—	—	—	—	±3.3
	July	—24.8	—38.8	—1.5	+13.6	—	—	—	—	±3.0
	August	—25.0	—39.1	—1.5	+13.7	—	—	—	—	±3.0
	September	—24.8	—39.5	—1.5	+13.6	—	—	—	—	±3.0
	October	—24.4	—33.6	—3.0	+15.6	—	—	—	—	±3.4
	November	—25.7	—33.8	—3.0	+16.0	—	—	+3.0	—	±3.5
	December	—30.8	—34.1	—3.1	+17.4	—	—	+1.8	—	±4.0
1981	January	—31.6	—30.1	—3.1	+17.6	—	—	+2.0	—	±4.0
	February	—50.0	—30.4	—3.1	+25.0	—	—	+2.2	—	±5.7
	March	—69.9	—30.6	+0.6	+27.0	—	—	+3.4	—	±6.4
	April	—67.0	—30.6	+1.1	+27.3	—	—	—	—	±6.3
	May	—67.2	—30.9	+1.1	+27.5	—	—	—	—	±6.3
	June	—68.1	—31.2	+1.1	+27.7	—	—	—	—	±6.4
	July	—71.7	—31.5	+1.6	+29.2	—	—	—	—	±6.8
	August	—72.3	—31.8	+2.2	+28.6	—	—	—	—	±6.9
	September	—72.9	—36.4	+2.3	+28.9	—	—	—	—	±7.0

Marshall Plan, certain participants in the EMS negotiations who thought that they would be forced into unacceptable deflationary policies because of the coexistence of their potentially weak currencies with potentially strong currencies, secured considerable credit facilities which were not touched.

Despite the categorical promise of the Brussels Resolution of December 1978, the question of whether and when these shortcomings will be remedied must be left to the future. Indeed the circumstances that have prevailed since the date originally set (March 1981) for the consolidation "of the scheme into a final system..." have not permitted the proposed action to take place and the system, although it has functioned with surprising ease so far, has suffered severe shocks during the year 1981. Further shocks may be felt because the exchange constellation which prevailed for most of the period ending between the two realignments of September 1979 and October 1981 have been reversed. In the meantime, while the EMS has not sufficiently promoted internal convergence between the economies of the countries concerned, it has so far protected them from the upheavals which have affected exchange rates outside the EMS.

3. Gold and the Future of the EMS

3.1 *Prospects for Change in Present EMS Mechanics*

The intentions of the authorities concerning the EMS for the near future are not known with precision. Although no official document has yet been published on the last November Summit meeting in London, press conferences have led to expectations that the change that may be decided upon will be of a rather technical nature.

It may therefore be surmised that the present treatment of gold will not be altered. This opinion is based on the fact that the ownership of gold — one of the more important questions to be settled at some stage — has not been transferred to a Community organization and that for the time being the creation of ECUs against deposits in gold (and dollars) is based on revolving swaps of three months duration, which may be unwound under certain rather loose conditions. In case of liquidation outstanding positions will be settled but debtor central banks will not be under an obligation to deliver gold.

Since it is not possible to discuss the authorities' intention regarding gold, an attempt will be made below to examine certain of the theoretical possibilities of treating gold in a reformed EMS.

3.2 *Theoretical Possibilities of Treating Gold in a Future EMS*

The imperfections of the present mechanics of the EMS have been mentioned above. They may be recapitulated as follow:

a) the arrangements for creating the ECU are precarious and the transfer of reserves is provisional; hence the limited "moneyness" of the ECU;

b) the volume of ECU creation varies in an automatic manner and is subject to the whims of the market price of gold (and dollars);

c) the European Fund for Monetary Cooperation is hardly anything but an accounting framework and has no real policy functions; ad hoc arrangements concerning relations between participating countries or between them and third countries take place bilaterally or multilaterally outside institutional dispositions.

If the EMS is to be transformed into a "final" form as suggested both by the Bremen communiqué and the Brussels Resolution, the transfer of reserves to the future "European Monetary Fund" will have to be made mainly with a view to:

a) establishing permanent arrangements for ECU creation;

b) enabling the Fund to devise and implement policies for contributing to financing and adjustment both within the EMS and between it and third countries.

While the maintenance of gold in the system would not be indispensible to the achievement of these objectives and the complexity of arrangements to that effect might prove to be a deterrent, it is not probable that gold be entirely kept out because the idea that the construction would need some "backing" is likely to endure. Accordingly, the theoretical possibilities set forth below assume that some role will be given to gold. These possibilities are presented as prototypes rather than mutually exclusive or exhaustive solutions.

A) Maintaining the swap system with certain modifications. The price of gold could be "frozen" (but it is open to question whether this would be in conformity with the reformed IMF Articles of Agreement); or, the amount of ECUs outstanding could be kept constant, in accordance with a "closed pool" scheme with or without adjustment through the volume of reserves paid into the system.

Since the very notion of swap implies a transitory situation, no mechanism based upon an arrangement of this type could make it possible to create a "permanent" ECU.

B) Definitive sale of reserves tranferred to the Fund. The gold (and dollar) reserves could be definitively surrendered to the Fund which would have full control over them.

By a sale, the central banks would definitively surrender a fraction of their reserves to the EMF and they would naturally expect to receive ECUs which could have such characteristics as to make them closely comparable to the reserves sold to the Fund, from the standpoint of convertibility and liquidity.

Moreover, it is likely that the present formula for the deposit of reserves (the 20 percent rule for gold and for dollars) would have to be modified, since the proportion of gold and dollars actually contributed to the Fund's assets would vary between central banks and would pose intractable problems of risk sharing. The transfer of assets in equal proportions (e.g. 2/3 gold and 1/3 dollars) might provide a solution, but unless the gold proportions were much lower than today some members might have to purchase gold.

C) Ad hoc formulae based on the specific character of gold assets. Given the specific problems linked to the transfer of gold, alternative options for transferring gold to the Fund which leave the transferrer of gold with rights over this transfer might be chosen. In such options, dollar transfer would take the form of a surrender pure and simple, while the treatment of gold might take the form of a:

(i) Transfer as capital.

The transferrer of gold would receive a title of ownership in the Fund's capital. The Fund could allocate simultaneously to the participating central banks an amount of money in ECUs proportionate to their capital participation. Participants would retain a right of recovery over the assets transferred to the Fund (in case of liquidation) and of benefiting from the profits from any revaluation. The Fund would not necessarily be precluded from selling the gold so as to preserve the convertibility of the ECU, and recovery rights as well as the limits of the Fund's possibilities of action would have to be stipulated in detail.

(ii) Transfer against issues of mobilizable certificates of deposits.

The gold transfer would not be definitive and the transferrer would receive a certificate of deposit, expressed in gold weight. For the certificate to be converted in an instrument of settlement — the ECU — it would have to be mobilized, at the holder's request, through the Fund (or through another central bank). This solution could mean that the creation of ECUs against gold would keep pace with the mobilization of the certificates; as a result, the volume of ECUs created at the start of the phase would be smaller than the amount in circulation at the end of the present transitional period.

(iii) Transfer of gold with preservation of rights over revaluation profits.

Under this option, the transfer of gold against ECUs would be accompanied by the preservation of a right to the profit from any revaluation of the gold. Since the Fund would pay interest on all ECUs held by the participants, and since gold would not produce interest, each member country would have to pay back compensatory interest on its share of ECUs received against gold.

Like the previous formula, this form of transfer would not amount to a definitive transfer; it would pose scarcely any accounting problems. Nor would it show up a reduction in the amount of ECUs outstanding at the time of transition to the institutional phase.

3.3 *Including Gold in the ECU Basket*

The idea that a monetary instrument should have some "backing" is still so strong that the inclusion of some gold in the composition of the ECU might be conceived as likely to improve its "moneyness" or at least to enhance its attractiveness as a reserve asset.

There are no technical problems in constructing a basket which would include gold as well as currencies. Had it been decided in March 1979 to start the EMS with a different composition of the ECU, one in which there was, say, 20 percent gold and each of the currency weights was reduced proportionately, that version of the ECU would include just over one hundredth of an ounce of gold.

As a technical explanation and the tables presented in the Appendix show, even such a small amount of gold would have caused considerable variation in the value of the ECU. Indeed, as the market price of gold has fluctuated widely over the period, an ECU which included gold would have been much more volatile in terms of any of the composing currencies as well as of any outside currencies than the actual ECU. For example the DM has appreciated against the ECU by about 2 1/2 percent since the start of the system, but if gold had been included it would have depreciated by nearly 40 percent. This whole purpose of using a composite currency (hedging against fluctuation in the value of individual currencies) would thus be ruined.

The market rate equivalent of an ECU which contained gold could be found using either the daily market rate as is done for the currencies, or the rate used in the swap operations (described in section 2.2.B (ii)). These two rates can differ substantially. On the market, the ECU price of gold went above 650 per ounce, but the swap price did not reach 450 ECU per ounce. Sometimes the effect can be in the opposite direction: the swap price can be *above* the market price.

At realignments it would be necessary to determine a "fictitious central rate" for gold in the same way as is necessary for sterling. This could be interpreted by some as giving a monetary price to gold which might be considered as contrary to the Second Amedment of the IMF Articles.

Fluctuation of gold's market price away from its fictitious central rate would cause the familiar (from sterling's example) problems for the divergence indicator (and for the agri-monetary system).

Most of the above difficulties are accentuated by gold having had such a variable price in terms of all currencies. Had the dollar been included in the ECU rather than gold, the problems would have been less because the dollar has varied much less against the basket currencies than has gold. (Against an average of them: its low point was only less than 10 percent below its starting value and the high point only about 40 percent above the low point. Gold's high point was nearly 20 percent above the starting point and its current price is nearly 50 percent below the high point). Although the price has risen substantially since March 1979 these fluctuations make gold a dubious store of value.

The inescapable conclusion from the preceding explanations is that the inclusion of gold in the ECU basket is not a practical proposition.

Whether the use of the ECU as a money in private banking and financial use will develop (such use has not only started but is developing) depends on others factors which will not be examined here.

4. Conclusions

4.1 *On the EMS and Gold*

The present paper has discussed the role of gold in the EMS. In the present transitional stage of that system that role is limited to the process of ECU creation through the deposit of 20 percent of the gold reserves of participating countries valued according to a method related to the market price of gold. This has given rise to considerable changes in the liquidity thus created and such changes are totally unrelated to economic criteria. Moreover they have provided participating countries with settlement possibilities which have hardly been touched. The properties of the ECU and the rules regarding its utilization, as provided for by an "Agreement between the Central Banks of the Member States of the EEC laying down the operating procedures of the EMS", are such that the attractiveness of this instrument has been extremely limited. In general, central banks have preferred to use the dollar for intra-marginal interventions which have been considerably more important than interventions at the limits. Access to credit provided by the international capital markets, which are rather unobtrusive and unconditional, explains this preference. There has, apparently, been no attempt on the part of the creditors to impose limitations on this practice, and coordinations of policies via the conditionality of the credit system could only be achieved if debtor countries would cease to have relatively easy access to capital markets, which has not happened so far. Thus, while existing EMS arrangements are more than a simple exchange rate system, they hardly amount to a monetary system, mainly because there is as yet no policy-making Fund.

The lessons taught by recourse to gold in its ancillary functions in the EMS are clear. The process of ECU creation through gold (and dollar) deposits valued at market price is cumbersome and devoid of economic justification; nor is it clear how these deficits could be remedied except through arrangements that are so complex that it would be better to keep gold (and dollars) completely out of a reformed system and let the authorities regulate its liquidity via liquidity creation based on their own decision. Last, the idea that the attractiveness of the ECU as a reserve asset could be improved by the inclusion of some gold in the basket has been shown to be completely impracticable.

4.2. *On the Wider Framework*

Other papers to be presented to this Conference will discuss the pros and cons of a recourse to gold in future international (as distinct from Community) arrangements. To the author of the present paper expressions of interest for a monetary role to be played by gold seem to belong to the field of articles of faith rather than to economic logic. The moral argument that central banks which have legitimately acquired gold assets in the past should be able to use them has ceased to have the validity it had at the time when sales to and purchases from the market were banned. President Zijlstra[9] in his plea "to consider ways to regulate the price of gold" admits that "step by step, not always painlessly, the freedom of central banks to effect transactions in gold, whether among themselves or in the free market, has been restored". Certain central banks have made ample use of this freedom[10] but most monetary and financial authorities have preferred to let their gold assets remain untouched while occasionally borrowing heavily, at high interest rates, from the market.

Perhaps a fitting conclusion of the present paper would be a remark that could be put in a form of words like: "Innocence once lost cannot be regained". There was a time when the world was innocent enough to accept "the role and the rule of gold". Even as late as 1931, Sydney Webb who had been a member of the previous Labour government, exclaimed, upon hearing the news that the link between sterling and gold had been cut: "They never told us we could do that!" Were it not for the conservatism inherent in

---

[9] This is the impression derived from a comparison between the plea of President Zijlstra in favour of "ways to regulate the price of gold" and the closely reasoned arguments of a number of experts who have expressed opposite views on this question. See for instance:
Edward M. Bernstein: a) *Is a Return to the Gold Standard Feasible?* (EMB Papers n° 80/19 October 16, 1980). b) *What Role for Gold in the Monetary System?* (EMB Paper n°81/21 November 12, 1981).
Dr. Schlesinger, Deputy Governor of the Bundesbank: Speech delivered at the University of Göttingen, BIS Press Review, n° 233 (Basle, December 1, 1981).
Robert Solomon: *Gold in the International Monetary System*, Address before the New York Society of Security Analysts.
Henry C. Wallich, Member Board of Governors of the Federal Reserve System: *Gold and the Dollar*, Remarks at 7th International Conference FOREX, Nov. 23, 1981.

[10] See *Financial Times*, December 4, 1981, p. 17: Article by David Marsh and figures for 1980 for Canada, Indonesia and others in *International Financial Statistics*.

money matters "that" could, perhaps should, have been done decades earlier, for instance when telegraphic orders became a more usual way of making settlements than transporting gold ingots; and perhaps also when universal suffrage became the alpha and the omega of political democracy. We are now turning a new page in that book; this new page is labelled "economic democracy" and its advent is being precipitated by the present powerful thrust towards egalitarianism. Innocent Sydney Webbs are gone forever, but working out a reconciliation between monetary stability and economic democracy remains an elusive objective.

5. Appendix: the Inclusion of Gold in an ECU Basket

5.1 *Construction of a Basket Containing Gold as well as National Currencies*

A basket, called gold-ECU hereafter containing 20 percent gold on 13 March 1979 was constructed by decreasing each of the fixed currency amounts in the actual ECU by 20 percent and adding in just over one hundredth of an ounce of gold. The precise amount was found by finding the gold equivalent of 20 percent of the dollar value of the ECU using a gold price of 221 dollars per ounce.[11] The fixed amounts in this "gold-ECU" basket are given in column 2 of Table 5; columns 4 and 5 show that the dollar equivalent of the two baskets, the "gold-ECU" and the actual ECU, can be found in the same way; and that on 13 March 1979 the values were equal.

5.2 *Subsequent Values*

Table 6 gives quarterly exchange rates for the basket currencies and gold. For gold the London morning fixing rate of the day was used. These rates can be used to calculate the "gold-ECU" equivalents in the way shown in Table 5. The results in dollars and DM are given with the equivalent amount for the actual ECU in Table 7.

5.3 *Composition of the Baskets.*

The dollar equivalent of the actual ECU and that of the "gold ECU" vary considerably over the period. The dollar first depreciated against the ECU and then appreciated strongly, with the maximum swing going from — 6.6 percent to 28.9 percent. Against the "gold-ECU" its depreciation was much greater, reaching a maximum of 30.9 percent. It subsequently nearly

[11] $(1.3516 \times 20/100) / 221 = 0.001223$

Table 5

	Initial composition of Actual ECU 1	Gold ECU 2	Exchange Rates 13.3.1979 3	Dollar Equivalent of actual ECU Col. 1 + Col. 3 4	Dollar Equivalent of Gold ECU Col. 2 + Col. 3 5
B/LF	3.80	3.04	29.4625	0.1290	0.1032
HFL	0.286	0.229	2.0096	0.1423	0.1140
DKR	0.217	0.174	5.2055	0.0417	0.0334
DM	0.828	0.662	1.8612	0.4449	0.3557
LIT	1.09	87.2	846.5	0.1288	0.1030
FF	1.15	0.92	4.2835	0.2684	0.2148
UKL	0.0885	0.0708	2.046*	0.1810	0.1449
IRL	0.00759	0.00607	2.046*	0.0155	0.0124
GOLD	—	0.001223	221*	—	0.2702
				1.3516	1.3516

* 1 t = x dollars; 1 ounce = x dollars.

Table 6 - *Exchange rates against the US dollar*

	5/7 July 1979	1/10 Oct. 1979	2/1 Jan. 1980	1/4 April 1980	1/7 July 1980	1/10 Oct. 1980	5/2 Jan. 1981	1/4 April 1981	1/7 July 1981	1 Oct. 1981	1 Dec. 1981
B/LF	29.2900	28.1000	27.8625	31.6900	28.2025	28.8670	31.4375	34.3225	39.4150	38.0900	37.4525
HLF	2.0160	1.9286	1.8915	2.1610	1.9324	1.9570	2.1210	2.3220	2.6742	2.5860	2.4250
DKR	5.2660	5.0795	5.3385	6.1380	5.4710	5.5600	6.0050	6.6000	7.5565	7.3175	7.1220
DM	1.8276	1.7365	1.7155	1.9788	1.7640	1.8030	1.9517	2.0952	2.4055	2.3240	2.2150
LIT	821.5000	801.0000	801.2500	911.9500	841.5000	859.2500	928.5000	1045.2500	1198.5000	1183.5000	1188.7500
FF	4.2500	4.0815	4.0165	4.5580	4.0900	4.1820	4.5210	4.9445	5.7430	5.5710	5.5930
UKL	2.2370	2.1933	2.2290	2.1355	2.3585	2.3935	2.4035	2.2427	1.9125	1.8350	1.9480
IRL	2.0650	2.1390	2.1550	1.9040	2.1270	2.0810	1.9030	1.7410	1.5160	1.5675	1.6015
GOLD	286.7500	399.7500	559.0000	501.5000	656.0000	679.2500	592.5000	516.7500	421.2500	435.2500	405.0000

1 USd = x units of national currency except for UKL, IRL and GOLD.

Table 7 - *ECU and "gold ECU"**

	Dollar equivalent			
	ECU *(1)*	*% change* *(2)*	*"gold ECU"** *(3)*	*% change* *(4)*
13 March 1979	1.3516	(—)	1.3516	(—)
5 July 1979	1.3828	— 2.3	1.4569	— 7.2
1 Oct. 1979	1.4313	— 5.6	1.6335	—17.2
2 Jan. 1980	1.4469	— 6.6	1.8411	—26.5
1 Apr. 1980	1.2813	5.5	1.6383	—17.5
1 July 1980	1.4274	— 5.3	1.9441	—30.5
1 Oct. 1980	1.4055	— 3.8	1.9551	—30.9
5 Jan. 1981	1.3150	2.8	1.7766	—24.0
1 Apr. 1981	1.2105	11.7	1.6003	—15.5
1 July 1981	1.0482	28.9	1.3533	— 0.2
1 Oct. 1981	1.0691	26.4	1.3876	— 2.6
1 Dec. 1981	1.1055	22.3	1.4247	— 5.1

	DM equivalent			
	ECU *(5)*	*% change* *(6)*	*"gold ECU"** *(7)*	*% change* *(8)*
13 March 1979	2.5156	(—)	2.5156	(—)
5 July 1979	2.5272	—0.5	2.6626	— 5.5
1 Oct. 1979	2.4854	1.2	2.8368	—10.3
2 Jan. 1980	2.4821	1.3	3.1584	—20.3
1 Apr. 1980	2.5355	—0.8	3.2419	—22.4
1 July 1980	2.5179	—0.1	3.4294	—26.7
1 Oct. 1980	2.5341	—0.7	3.5250	—28.6
5 Jan. 1981	2.5665	—2.0	3.4674	—27.4
1 Apr. 1981	2.5363	—0.8	3.3529	—25.0
1 July 1981	2.5215	—0.2	3.2563	—22.8
1 Oct. 1981	2.4846	—1.2	3.2248	—22.0
1 Dec. 1981	2.4488	2.7	3.1557	—20.3

* See text for derivation.

regained its March 1979 position in July 1980 with the decrease in gold price and the depreciation of the European currencies. The DM equivalents of the ECU and the "gold ECU" show a marked difference. The variation of the DM against the actual ECU is very limited reflecting the relative stability of the Community currencies and the weight of the DM in the basket. Against the "gold ECU", the DM depreciated nearly as much as the dollar, but has not recovered.

CHAPTER 20

THE ROLE OF GOLD IN THE INTERNATIONAL MONETARY FUND TODAY

GUNTER WITTICH

1. Introduction

My topic at this Conference — The Role of Gold in the International Monetary Fund Today — could be dealt with in relatively few sentences. They would indicate that gold has been removed from the Fund's Articles as a basis for transactions, and that its role in the exchange system and as a numéraire has been eliminated. Such a summary would, however, fail to bring out the changes that have occurred over the years, not all of which were directly related to the Fund. Perhaps more importantly, it would neglect that gold remains an important asset among international reserves that has to be taken into account in the surveillance of international liquidity.

This morning, I would like briefly to recall the major changes in the role of gold in the monetary system, and to summarize the agreements which led to its present role. I will then outline how these agreements have been reflected in the Second Amendment of the Fund's Articles of Agreement, which entered into effect almost four years ago; and summarize sources and uses of the Fund's gold and its gold sales program. I will end with a few words on the difficult question of the valuation of gold in international reserves.

2. Changes in the Role of Gold in the International Monetary System

The decline of the importance of gold in the international monetary system occurred in stages, at times perhaps imperceptibly, and at other times with much commotion. It reflected in part the growing difficulties experienced with the fixed exchange rate system as it worked in practice under the Bretton Woods Agreements. These stages are all familiar to you, and it will suffice if I recall some of the major aspects and events.

First, the growth in reserve assets in the form of currency holdings — mainly US dollars — that resulted in a substantial decline in the share of gold in the international reserves.

Second, the growing concern about the inherent instability of the gold exchange standard, first articulated by Professor Triffin in the late 1950's.

Third, the establishment in 1961 of the Gold Pool by a number of major industrial countries to stabilize the price of gold as a means of stabilizing the exchange rate system.

Fourth, the agreement, incorporated in the First Amendment of the Fund's Articles, to establish the Special Drawing Rights Account in the International Monetary Fund to meet the need, as and when it arose, for a supplement to existing reserve assets.

Fifth, the separation of the gold market into two tiers in 1968, by which private transactions in gold were separated from official transactions. In view of the prospective establishment of the facility for special drawing rights, it was felt that monetary authorities would no longer find it necessary to buy gold from the market.

Finally, the decision of the United States authorities in August 1971 to end officially the free convertibility of dollars held by monetary authorities into gold at the par value of the US dollar. This decision effectively ended the gold exchange standard of the postwar Bretton Woods monetary system.

3. The 1975 Consensus on the Future Role of Gold

These changes in the role of gold were, of course, only part of the broader developments in the international monetary system during the postwar period. These developments, which ended in the breakdown of the par value system, led to a protracted discussion of possible reforms of the system in the early 1970's, most prominently in the Committee of Twenty and in meetings of the Interim Committee. In this context a consensus eventually emerged on achieving a gradual reduction of the role of gold.

The main elements of this consensus were that there would no longer be an official price of gold; that the various obligations of member countries to use gold in transactions with the International Monetary Fund would be abolished; that a part of the Fund's gold holdings would be sold for the benefit of developing member countries; and that another part would be sold to all then Fund members at the official price.

It was also agreed that there would be no action to peg the price of gold. The total stock of gold in the hands of the Fund and of the Group of Ten countries and Switzerland would not be increased. The latter agreement entered into effect in February 1976 for a period of two years; in February 1978 it was allowed to lapse in view of the then pending Second Amendment of the Fund's Articles.

4. The Second Amendment of the Fund's Articles

In due course, these agreements found expression in the Second Amendment of the Fund's Articles. As gold formed such a pervasive element in the Fund's original Articles — which in this respect were not much changed by the First Amendment establishing Special Drawing Rights — the changes introduced by the Second Amendment were numerous.

The Second Amendment of the Articles, which entered into effect on April 1, 1978, removed gold from the centre of the exchange rate system. It abolished the official price of gold and the specifc regulation on transactions in gold among member countries — which had prohibited members from purchasing (or selling) gold at a price above (or below) its par value — but introduced a general obligation for members to collaborate with the Fund regarding international liquidity. It eliminated gold as obligatory means of payment between members and the Fund; and set out ways in which the Fund could utilize its remaining holdings of gold.

More specifically, the most important changes can be summarized as follows:

1) Gold was eliminated as the common demoninator of the par value system and as the unit of value of the SDR.

Before the Second Amendment, par values were defined in terms of gold or in terms of the US dollar with the weight and fineness in effect on July 1, 1944. There are now no par values, but the amended Articles do allow for the reintroduction of a system of exchange arrangements based on stable but adjustable par values. However, these par values would have to be expressed in terms of the SDR or some other common denominator prescribed by the Fund: as a legal matter, gold (or another currency) cannot be such a common denominator.

2) The official price of gold was abolished through the elimination of the definition of the value of the SDR in terms of gold, and the value of the Fund's holdings of members' currencies are now maintained, and its transactions denominated, in terms of the SDR.

The elimination of the official price of gold also affects dealing in gold among member countries. As I mentioned earlier, before the Second Amendment monetary authorities were not permitted to purchase gold at a price higher than that corresponding to the par value, or sell it at a lower price. In effect while market prices differed from the official price, gold was unlikely to be used as a means of payment between monetary authorities, which generally could not be expected to sell gold at a price lower than the price in the private market. Under the amended Articles, there are no restrictions on the freedom of monetary authorites to enter into gold transactions among themselves or with the market, except for the general undertaking regarding world liquidity and the role of the SDR.

The deletion of the definition of the SDR in terms of gold and authorization of the Fund to determine the method of its valuation completed the change in the valuation of the SDR adopted in 1974 in response to the inconvertibility of currencies into gold. This inconvertibility had put into question the value of the SDR in terms of all currencies, which previously had been derived from the gold par value of one currency. Inconvertibility had left undecided the question whether any, and if so, which members might be considered still observing gold par value; and the value of the SDR had become dependent on the currency used as the link. As you know, this dilemma has been dealt with by the introduction of a basket valuation for the SDR, which in turn then defined the official price of gold in terms of currencies.

Initially, the basket valuation of the SDR was based on 16 currencies, and started with the then existing value of the SDR in terms of US dollars. The change in the valuation method thus did not imply a general increase in its value (or decrease in the value of currencies).

In the meantime, the Fund has made use of the authority to determine the value of the SDR, which today is based on a basket of the currencies of the five members in the Fund with the largest share in international trade and payments.

3) The Fund was forbidden to re-establish a fixed price of gold. In its dealing in gold the Fund is required to avoid management of the price in the market, or the establishment of a fixed price.

4) The obligatory use of gold as a means of payment to and from the Fund was eliminated; and any use of it in payments requires a very high majority of voting power and thus wide agreement among the Fund's membership. These obligatory payments included the payment of a part of the subscription when a country joined the Fund or when quotas were increased (which are now made in SDRs unless the Fund determines otherwise), certain repayments of indebtedness to the Fund, and the payment of charges. In some cases, members in the past had no option but to make payments in gold; in others, the option to use gold for payments rested with the member and the Fund had no right to refuse such payments.

Any use of gold in the future would be voluntary on both sides, and would be based on prices agreed for each transaction on the basis of prices obtaining in the market. However, acceptance by the Fund of gold in payment requires approval by 85 percent of the voting power in the Fund's Board of Executive Directors.

5) The use of the Fund's remaining gold holdings was regulated. As I have mentioned earlier, the agreement on the future role of gold included a decision to sell part of the Fund's gold holdings. These sales had started

before the Second Amendment of the Articles entered into force, and the Amendment instructed the Fund to complete them. For additional gold sales, the Amendment required a high majority of 85 percent of the voting power. Subject to a decision taken with this majority, the Fund can sell from its remaining gold holdings at the official price to countries that were members of the Fund on August 31, 1975 in proportion to their quotas on that date, or at a price based on prices in the market to members or others. The Amendment also specified in considerable detail the use to which sales proceeds in excess of the former official price can be put. These uses include temporary income-producing investments; assistance to developing members in difficult circumstances; and addition to the Fund's general resources for use in its normal operations, accompanied by a proportionate increase in the quotas of countries that were members in August, 1975.

6) As I have mentioned, the Amendment prohibited the Fund from re-establishing a fixed price for gold in its dealings. The concern that dealing in gold by the Fund might establish or be seen to establish a price for gold gave rise to the circumscription that the Articles place on the pricing of gold in any sales by the Fund. Gold sales — or, for that matter, purchases if they were agreed — are to be made "at prices based on the market price"rather than "at the market price": this is intended to specify that in its transactions in gold the Fund should seek to follow rather than to determine or set a direction for prices in the markets. But the Amendment went one step further when removing gold as the basis of the international monetary system. The amended Articles included an undertaking by members to collaborate with the Fund and with other members with respect to reserve assets, so as to achieve better surveillance of international liquidity, and to promote the role of the SDR as the principal reserve asset in the international monetary system.

5. Gold in the Fund's Financial Structure and the Gold Sales Program

Let me now say a few words on the Fund's gold holdings and the gold sales program. At the time of the inception of the Fund, gold formed the ultimate reserve asset, and it is not surprising that it was considered important for the financial strength of the institution that part of the subscription be payable in gold. The original Articles thus contained a provision that part of its assets be paid in gold. As mentioned earlier, a number of other payments, such as charges or specified repayments to the Fund, had to be made in gold. The Fund also acquired gold in a variety of the other ways, including purchases from South Africa in the context of the two tier gold market arrangement.

The Fund's gold holdings in turn were used to support its financial activities. Gold was sold to replenish the Fund's holdings of currencies, for example, when large drawings took place that also necessitated borrowing under the General Agreement to Borrow (GAB). Gold was sold in connection with quota increases in order to facilitate the gold payments required for the increased subscriptions; and gold acquired from South Africa after establishment of the two tier market was sold to members to assist in meeting their asset preferences following the separation of official from private markets.

At the time of the discussions of the reform of the monetary system and the future role of gold in that system, the Fund's gold holdings were equal to SDR 5.3 billion (more than 150 million ounces or about 4,710 tons); the Fund gold stock was the second largest of official holdings. As part of the consensus on the reform of the monetary system and as a contribution towards reducing the role of gold, it was decided in 1975 that one third of the Fund's holdings or 50 million ounces should be sold. One sixth, or 25 million ounces were to be sold to the market for the benefit of developing countries. The profits on these sales — that is the proceeds in excess of SDR 35 an ounce — would be channeled to a Trust Fund established and administered by the Fund. Profits corresponding to the share of Fund quotas of developing member countries would be transferred directly to these members, and the remainder, corresponding to the share of Fund quotas of industrialized countries, would be used to provide balance of payments loans on concessional terms to the least developed of developed countries.

Sale of the other 25 million ounces, somewhat misleadingly called "restitutions", was to be made at SDR 35 an ounce to all countries that were members of the Fund in proportion to their Fund quotas on August 31, 1975.

The Fund's gold sales were carried out over the four years from June 1976 to May 1980, during most of which gold prices continued to rise. In order to avoid any appearances of the Fund seeking to manage the price of gold in the private market or of it taking a view on future price developments, gold was sold through a series of public auctions. Initially auctions were held at irregular intervals of six to eight weeks; from March 1977 on, they took place on a regular monthly basis to reduce uncertainty as to the supplies coming on to the market from the Fund.

In all, the Fund held 45 auctions, in which it sold 23.5 million ounces on competitive bids to buyers in the private markets, and 1.5 million ounces on non-competitive bids to monetary authorities. These latter sales represented the share in gold profits that some central banks preferred to take up in the form of bullion rather than cash payments, which they were entitled to do, after the Second Amendment, under the terms of the gold sales program. Prices in the auctions ranged from US $108.76 an ounce to US $718.01 and averaged US $228.56 an ounce, very close to the average of London fixing prices during the four-year period of US $234.52. Proceeds of these gold sales amounted to US $5.7 billion, of which US $1.1 billion represented the capital value of SDR 35 an ounce which was added to the Fund's general resources, and US $4.6 billion represented profits that were channeled to the Trust Fund. Of this latter amount, US $1.3 billion was distributed directly to developing member countries and the remainder, together with investment income on undistributed profits and other transfers to the Trust Fund, made available as loans to eligible members for balance of payments assistance on concessionary terms.

Gold sales to member countries at the official price in relation to their quotas ("restitution") were completed in annual installments over the same four-year period.

6. Possible Further Use of the Fund's Gold

The Fund's gold sales program was completed in May 1980, and its present gold holdings amount to about 103 million ounces.

These holdings contribute to the Fund's financial strength and to its ability to borrow additional resources to meet its members' needs. As mentioned, any decision on further gold sales by the Fund requires a majority of 85 percent in its Executive Board and thus presupposes very wide agreement among its members that such sales would contribute towards the achievement of the Fund's purposes.

One proposal regarding the Fund's remaining holdings was made at the time of the discussion of a possible Substitution Account. It was suggested that some gold should be set aside for the purpose of strengthening the financial structure of such an account. The proposed Substitution Account was designed to promote a more stable international monetary system by enabling member countries voluntarily to exchange official reserves of US dollars (and later perhaps also other currencies) for SDR-denominated claims on the Substitution Account. As reserves held in the Account would not be mobilized for shifts between differently denominated reserve assets, pressures on exchange rates deriving from the diversification of reserve porfolios would be reduced.

US dollars paid into the Substitution Account would have been invested in US government securities, while SDR claims would have paid the SDR rate of interest. One of the difficulties in designing the account arose from the likely difference in interest earned and interest paid, and in changes in the capital value of its assets and liabilities. While it might be expected that over the longer term interest and exchange rate differentials would be offsetting, this was not likely to be the case at all times. It was to meet such contingencies that consideration was given — but not agreed — to provide a cushion by the use of part of the Fund's remaining gold holdings.

7. The Valuation of Gold and Its Role in World Liquidity

I have mentioned earlier that gold still forms a substantial part of international reserve holdings of a number of central banks. The importance of its share in individual country's reserve holdings varies greatly as gold holdings are highly concentrated. Almost four-fifths of official stocks are held by the countries of the Group of Ten; holdings of other countries (with some exceptions such as South Africa) are generally small in absolute amount and in relation to other forms of international reserves.

How important that role is, however, remains a matter for conjecture and it is difficult to attach any exact value to the gold holdings of monetary authorities. Two factors in particular have introduced a substantial element of uncertainty in the valuation of gold: the abolition of a fixed and guaranteed price at which monetary authorities are assured of a possibility to buy or sell gold if they so wished, and the comparative narrowness of the private gold market and consequent potentially sharp fluctuation in the market price.

There is no longer the certainty that monetary authorities can mobilize their gold reserve for settlement or intervention purposes at a known price. While these reserves can be used as collateral for loans from other monetary authorities or from commercial banks, the value that will be attached to them is not known in advance. Normally it will depend on the price of gold in the private market and thus is influenced by the factors that affect the day-to-day interplay of supply and demand in a commodity market. Moreover, the authorities would face the risk of adverse price movements should they find it necessary to sell substantial quantities of gold in these markets.

It is therefore not surprising that a great diversity in the valuation of gold in official reserves has resulted. Of the countries reporting official gold holdings to the Fund for publication in *International Financial Statistics,* 56 continue to account for their holdings at the former official price. These countries hold about 48 percent of the official gold reserve of Fund member countries and Switzerland. Thirty-five countries value gold holdings at market-related prices: 22 of them use the latest market price or an average of prices over a recent period, and 13 use market or average prices reduced by a discount which ranges from 5 percent to 35 percent. The remaining 18 countries use a variety of other methods to set the value of their gold holdings.

Finally, members of the European Communities have deposited 20 percent of their gold holdings and of their foreign exchange reserves with the European Monetary Cooperation Fund, for which they have received an equivalent amount of ECUs. These deposits are renewed quarterly and the amount of ECUs corresponding to the gold deposited is determined on the basis of the average of London gold fixing prices during the preceding six calendar months, but not exceeding the average of the two London fixings on the penultimate day of the period.

For the Fund, the question of valuation of gold arises in a number of contexts. The amended Articles prescribe the manner of valuation of gold for some purpose, while in other contexts the choice of method is left to the discretion of the Fund. The method of valuation of gold under the amended Articles is thus not a simple or unitary one, but rather allows for the application of different principles as the context may require.

Just a few examples. The Fund must, of course, value its present holdings for presentation in its financial statements. For this purpose, the Fund continues to value its holdings at SDR 35 an ounce, which is also the price of acquisition.

The need to value gold stocks also arises in the context of the Fund's financial operations with member countries. When the Fund extends loans in support of stabilization programs, it primarily uses the currencies subscribed by its member countries. The particular currencies to be used during given periods are determined on the basis of balance of payments strength and reserve holdings of member countries. In this context, gold has of course to be valued consistently, and independently of the valuation attached to it by the member itself: for this purpose, also, SDR 35 an ounce is used.

One more example where the valuation of gold is of importance for the Fund is in the appraisal of the adequacy of international liquidity. Such appraisals are undertaken periodically and published in the Fund's Annual Reports. They are also necessary in conjunction with decisions to allocate SDRs and with quinquennial reviews of Fund quotas. Here the valuation of gold presents particularly difficult issues, and rather than attempting to establish any general guidelines, the method of valuation has been handled on an ad hoc basis depending on the analysis. It has often been found useful to present more than one measure. In its Annual Reports, for example, the Fund in recent years has presented data on gold holdings both in physical terms or at the constant value of SDR 35 an ounce and at market prices.

The valuation of gold in international reserves, whether for individual countries or in the aggregate, thus clearly has become more complex. It is also an area which is of great importance not only to individual countries, whether or not gold forms an important part of their reserves, but to the system as a whole. It is for this reason that one might expect it to be a question that will continue to be of concern to the Fund's member countries in order to promote a better international surveillance of international liquidity.

8. Conclusions

Let me end with two or three broad conclusions about the role of gold in the Fund today.

Gold no longer serves in the exchange rate system as a numéraire for fixing relative currency values; and the Fund is not to manage the price of gold.

Gold has been effectively eliminated for regular transaction and operations between the Fund and its member countries. It no longer serves as the ultimate source of the Fund's liquidity. However, as for a number of members, gold remains an important part of the Fund's assets and thus its financial strength.

Further use by the Fund of its gold holdings is comprehensively regulated by the amended Articles. Any sales of gold call for a wide consensus among the Fund's membership so that gold sales will serve the purposes of the Fund and the community of countries that form the international monetary system.

CHAPTER 21

THE PAST, CURRENT AND PROSPECTIVE ROLE OF GOLD IN THE US MONETARY SYSTEM

ANNA J. SCHWARTZ

My remarks today are based on a review of US experience with gold from 1834 to date that I prepared for the US Gold Commission. After noting the chronological changes in US experience, I summarize the evidence on the stability of real output and the price level during the periods when the United States was on or off the gold standard. I then describe the current role of gold in the United States, and conclude with a statement of proposals that have been suggested to increase the role of gold in the United States, if not also in the rest of the world.

1. Chronological Changes in US Experience with Gold Since 1834

From 1834 to 1973, with the exception of the years 1862 through 1878 and of an interlude of less than a year's duration in 1933-34, the United States adhered to some form of a gold standard. Chronologically, US experience with the gold standard may be characterized as follows:

1) 1834-1861: a de facto gold standard in a largely bimetallic international monetary system
2) 1862-1878: the greenback standard
3) 1879-1914: a gold standard without a central bank, and a fractional reserve banking system, as part of an expanding international gold standard
4) 1914-1933: a managed gold standard, under the Federal Reserve System, which was legally obligated to maintain minimum gold reserves against its monetary liabilities, in a short-lived postwar international gold exchange standard
5) 1933-1934: a floating dollar in an international monetary system split between a depreciated sterling area and a gold bloc clinging to parity

6) 1934-1948: the interwar and World War II and immediate postwar managed gold standard, in a fragmented international monetary system
7) 1948-1968: the Bretton Woods dollar/gold standard system, with progressive dilution of the gold restraints on US monetary conduct
8) 1968-1973: the breakdown of the Bretton Woods system
9) 1973-1981: The United States on an inconvertible paper dollar standard.

2. A Summary of Evidence on the Stability of the Economy When the United States Adhered or Did Not Adhere to a Gold Standard

There were only two extended interruptions in United States adherence to some form of a gold standard since 1834. One was 17 years in duration from the Civil War period of suspension of specie payments and the postwar period until resumption on January 1, 1879. The other interruption occurred in this century, for 13 years, if one dates the interruption from 1968, when the two tier London gold market was created; for 10 years, if one dates it from 1971, when convertibility of the dollar, even for official transactions, was formally suspended; for 8 years, if one dates it from 1973, when floating exchange rates were formally adopted by the United States and the Western industrial countries. The political objective of returning to the gold standard was achieved in the 19th century case, despite opposition from silver and paper money advocates. Whether that political objective is currently achievable cannot be determined from a retrospective view.

In addition to the two extended interruptions in US adherence to a gold standard, temporary suspension of a few weeks to a year's duration occurred in 1837, 1839, 1857, 1893, 1907, 1917-19, and 1933. In all cases but the latter two, the years in question climaxed periods of economic expansion in the United States, fostered by external as well as internal factors. The pace of the expansions raised US price and incomes above those prevailing in the rest of the gold standard world. To bring the US price and income structure into alignment with that of its trading partners enforced reductions in the US money stock, usually resulting from a decline in US gold reserves and in capital imports from abroad. Prices, output, and employment subsequently declined, accompanied by bankruptcies of firms and bank failures. Suspension of specie payments in the years under review was a means of mitigating the costs of deflationary adjustment that maintaining par values of the exchange rate imposed. The devaluation implicit in suspension gave the economy a breathing spell. With recovery, the former par value of the exchange rate was restored.

No special comment is needed on the World War I restriction of interconvertibility between paper money and gold and of the free international movement of gold. The situation in 1933, however, does require comment. That year was in no respect similar to the earlier examples of temporary devaluations. 1933 was a year of a business cycle trough after four years of deflation. The deliberate reduction in the gold content of the dollar was arranged to achieve a price rise of non-gold commodities, and the devaluation was never reversed. Moreover, the fixed exchange rate gold standard to which the United States returned in 1934 was the same in name only as the pre-1933 gold standard.

Before 1914, gold flows in and out of the United States determined the expansion or contraction of the economy. Between 1919 and 1933, large outflows of gold occasioned contractionary actions by the monetary authorities: small outflows and inflows of gold, whether large or small, were sterilized. After 1934, both inflows and outflows were not permitted to determine monetary growth and the performance of the economy. When the gold reserve ratios applicable to Federal Reserves deposit and notes were close to the minimum legal requirement, the minimum was lowered and eventually abolished. Gold became a symbol rather than an effective constraint on the operation of the monetary authorities.

Evidence on the performance of the economy is provided by the behaviour of a wholesale price series from 1834 on, with some reference also to the price series implicit in net national product, and of a real per capita income series from 1869 on.

Trend movements are the most striking feature of the price data. From 1834 to 1861, a mild downward trend prevailed, with pronounced cyclical upswings and downswings around the trend. The greenback period from 1862 to 1878 shows the sharp wartime price rise to 1865 followed by a decline of equal magnitude spread over the years to the close of the period. That decline persisted during the gold standard period of 1896, reflecting the disparity between the rate of growth of the monetary gold stock and the enlarged world demand. The reversal of the downward trend from 1896 to 1914 reflects the dramatic increase in world gold output during that period. World War I, like the Civil War period, shows a steep price increase to 1920, followed by the steep price decline from 1920 to 1921, rough stability during the 1920's, and then the great deflation of 1929-33 that restore the wholesale price series to its pre-World War I level, the implicit price deflator to a somewhat higher point than the pre-World War I level. The contraction of 1937-38 is apparent in the post-1933 upswing which continues into and beyond World War II. The wholesale price series shows rough stability in the early 1960's, whereas the implicit price deflator continues an upward movement. Both series accelerate after the mid-1960's.

If we examine annual estimates of real per capita output it is evident that the trend has been strongly positive from 1869 to 1980, as might be expected. There was substantial variance about the trend before 1914 but far smaller in magnitude than from 1914 to 1947, reflecting the sharp swings in the three interwar deep depressions, 1920-21, 1929-33, 1937-38, as well as the wartime movements. However, the pre-World War I variance was marginally greater than the variance of the deviations from trend post-1948. A comparison of the standard deviations of year-to-year percentage change in real per capita income also shows little difference between the pre-World War I gold standard experience and post-World War II experience: 5.8 percent vs. 5.5 percent. Unemployment was on the average lower in the pre-1914 period than in the post-World War I period; 6.8 percent vs. 7.5 percent. But again, excluding the interwar years, unemployment 1946-80 averaged 4.8 percent, reflecting the government's commitment to maintaining employment.

If we compare the purchasing power of gold, derived in index form from the quotient of the price of gold divided by the wholesale price index, with the US monetary gold stock, we observe that under the gold standard, a rise in the purchasing power of gold ultimately increased the growth of the US monetary gold stock by raising the rate of world gold output, and inducing a shift from non-monetary to monetary use of gold. Movements in the purchasing power of gold thus preceded long term movements in the monetary gold stock. This relationship underlay the reversion of the price level towards stability under the gold standard. Price increases or decreases tended to be reversed after a run of years. Persistent inflation of post-World War II experience, without a force to reverse the trend, could not have occurred under a fully functioning gold standard. The absence of this positive association after World War II between the purchasing power of gold and long term movements in the monetary gold stock reflects the loosening of the link between the money supply and the gold stock.

Over shorter periods, the relationship under the gold standard was in the opposite direction. Changes in the monetary gold stock, by influencing changes in the money supply, produced a negative association between the purchasing power of gold and the gold stock. Thus an increase in the gold stock would lead to an increase in the price level and, for a given nominal price of gold, lower the purchasing power of gold. The negative association may be observed during the gold standard period, changes in the monetary gold stock leading short term movements in the purchasing power of gold.

Finally, we may compare the exchange value of money, computed as the reciprocal of the wholesale price index, with the purchasing power of gold. The two series are closely related until 1968, when the two tier market for gold was introduced. The direct relationship until 1968 reflected the existence of a fixed nominal price of gold. The inverse relationship thereafter reflects the increase in private demand for gold as a hedge against inflation and political instability, once private transactions were determined in the free market.

To conclude: the gold standard provided long term but not short term price predictability. Long term inflation or deflation under the pre-World War I gold standard would predictably be reversed as gold output was discouraged or encouraged by decreases or increases in its purchasing power. Thus the price level tended to revert toward a long run stable value under the gold standard, providing a degree of predictability with respect to the value of money. Subsequent to World War I, the discipline of the gold standard came to be regarded as an impediment to the management of the economy to achieve the objective of growth and high employment. The deep depressions of the interwar years were the measure by which the economy under a gold constraint was judged to be a failure. The loosening of the link to gold after World War I and its abandonment fifty years later reduced long term price predictability. Belief in long term price stability eroded as public perception of the absence of a long run constraint on monetary growth took hold. Although price stability was generally included among the goals of the post-World War II era, in fact stability of employment took precedence. In any event, by 1981, neither goal was in sight.

3. The Current Role of Gold in the United States.

When pegged exchange rates were abandoned in March 1973, it was initially assumed that floating was a temporary expedient to be succeeded by a reformed par value system. The United States took the lead in opposing the return to such a system, on the ground that the dispersion of inflation rates among the industrialized countries and the higher variability of rates of inflation since the late 1960's enforced more frequent changes of exchange rates. Under the earlier system, changes in par values were delayed until foreign exchange market crises were provoked. The lesson since the shift in March 1973 was that floating provided more flexibility. The US view prevailed.

After the float, the US took the position that gold should be demonetized. At issue was the use of gold in official transactions at the free market price, and the substitution of gold for the dollar in inter-central bank settlements at a fixed but higher official price. In August 1975, agreement was reached by an IMF committee that the official price of gold would be abolished. The United States, however, valued its own gold assets at the official price of $42.22 per ounce, despite the IMF's abolition of that price.

The US repealed the prohibition against gold holding by US residents as of January 1, 1975, and empowered the Treasury to offset any increase in market price as a result of this increment to private demand by offering gold at auction. The first auctions were held in January and June 1975, when the Treasury disposed of 1.3 million ounces. No auctions were held in 1976 and 1977. They were resumed in 1978 and 1979, when the Treasury sold 4.0 and 11.8 million ounces, respectively, motivated as much by the desire to reduce the US balance of payments deficit on current account as by the belief "that neither gold nor any other commodity provides a suitable base for monetary arrangements."

The gold sales constituted open market operations approximating $0.8

billion in 1978 and $3.6 billion in 1979. Gold sales by the Treasury reduced the public's deposits and so bank reserves. The sales thus initially provided a partial offset to Federal Reserve open market purchases of government securities that increased the public's deposits and bank reserves. It was a partial offset only because the System's portfolio of government securities showed a net increase of $7.7 billion in 1978 and of $6.9 billion in 1979. It was an offset initially only depending on the Treasury's use of the proceeds of the gold sales. To the extent that the Treasury used the proceeds to retire gold certificate credits and thereby reduced its deposits at the Federal Reserve, the monetary effects of the gold sales were contractionary. However, to the extent that it disbursed the remainder of the funds it acquired, the Treasury's action restored the public's deposits and bank reserves, so the contractionary effect on the money supply of the gold sales was limited.

Since 1979, the Treasury has sold no gold bullion. In July 1980, however, it began the sale of half-ounce and one-ounce gold medallions, in accordance with P.L. 95-630, November 10, 1978. The legislation provided that not less than 1 million troy ounces of fine gold be struck into medallions and sold to the public over a five-year period at a price covering all costs. In 1981, US government gold inventories amounted to 264.2 million ounces. The Reagan administration has announced that its position on the proper role of gold in the domestic and international monetary system will not be formulated until the Congressionally mandated Gold Commission issues its report.

4. Proposals to Increase the Role of Gold in the United States

The creation of the US Gold Commission reflects the belief of many citizens that inflation in the United States can be curbed only by the adoption of some form of a gold standard. Some favour the restoration of the Bretton Woods system, some the form of the gold standard that prevailed in the United States between World War I and 1933. Ideologues dismiss the forms of the gold standard that have been known in the past as pseudo-gold standards and argue for a so-called real gold standard, in which national monetary units would not exist, coinage would be struck in weights of gold, and all prices would be expressed as weights of gold.

Links to gold other than a full-fledged gold standard are also being advocated. Some would involve the restoration of gold coins valued at a changing market price, possessing legal tender and convertible at the market price at the Treasury. It is claimed that such gold coinage would serve as a parallel currency to paper currency. How transactions would be conducted with a changing value of the medium of payment has not been carefully considered. Some would experiment with a limited issue of gold-backed bonds by the US Treasury partly to determine the premium or discount relative to bonds denominated in nominal dollars.

Another proposal would provide for a gold cover requirement for Federal Reserve note circulation and bank reserve. Convertibility of notes into gold is not contemplated. Some suggest upper and lower limits to the gold ratio. One proposal suggests raising the official price at which the gold stock is valued by a pre-announced percentage each year over a period of 8 years or so, to permit growth in the Federal Reserve monetary base. The gold reserve requirement would be a minimum one. Presumably, at the end of the period, the official price would be in line with the market price of gold.

Raising the price at which the United States currently values its gold stock is also proposed by those who advocate restoring the use of gold in the settlement of balance of payments disequilibria or for intervention purposes in foreign exchange markets.

My conclusion is that advocacy of a role for gold in the US domestic monetary system will increase if there is not convincing evidence during this year and 1983 that persistent inflation has been curbed. The silver issue dominated the US political scene in the last quarter of the nineteenth century. The gold issue may come to dominate the US political scene in the remaining decades of this century unless the managers of economic policy demonstrate their will to eliminate inflation.

CHAPTER 22

THE ROLE OF GOLD IN CENTRAL BANKS' RESERVE COMPOSITION

RAINER S. MASERA and GIANANDREA FALCHI

> ..."gold is a barbarous relic" (J.M. Keynes, *The Means to Prosperity,* 1933, reprinted in *Essays in Persuasion,* published for The Royal Economic Society, McMillan, 1972).
>
> "Gold still possesses great psychological value which is not being diminished by current events; and the desire to possess a gold reserve against unforeseen contingencies is likely to remain" (J.M. Keynes, "Proposals for an International Clearing Union", 1943, reprinted in H.G. Grubel, *World Monetary Reform*, Stanford University Press, 1963).

1. Introduction and Summary

The role of gold in the working of the international monetary system has undergone profound changes during the present century. Even with the explosion of foreign exchange holdings in the seventies and the fall in the price of gold during 1981, at the end of last year official gold valued at market prices still constituted the dominant component of official reserves, representing some 50 percent of total international liquidity. This, however, is sharply down with respect to the peak of over 90 percent reached in 1937, when gold could still be considered the world's international monetary base.

The first part of this note provides a brief review of the use of gold as an international reserve asset since 1913. To this purpose, summary statistics on gold production and absorption into monetary and non-monetary stocks are also presented. The interaction between volume and valuation changes is then analysed, with a view to assessing the important changes in the functions, value and composition of gold holdings of central banks.

The second part of the paper directly deals with developments in the postwar period and refers specifically to the joint questions of yield and its variability of today's principal reserve assets. Empirical elements are provided to assess the relative performance of gold and of the main currencies in the light of a multi-asset portfolio approach. To this end a backward simulation for the period 1950-1981 is made by comparing returns on gold with those of investments in dollars, pound sterling, yen, French francs, DM, and SDR s.

On the basis of these empirical findings some considerations are finally developed on the short term prospects of gold in the framework of today's multicurrency reserve system.

2. Gold in the International Monetary System: 1900-1981

The role of gold in the international monetary system in the present century can be reviewed by considering five distinct phases, which can be roughly separated according to intervals of some 20 years each.

2.1 In 1900 both national and international monetary systems were based on a fully-fledged gold standard. Great Britain had been the first major country to meet the formal conditions of the gold standard mechanism already in 1821, but many important countries adhered to the gold standard requirements only after 1895. The period 1850-1913 was characterized by an enormous economizing on gold in circulation resulting from transfers from the public to the reserve coffers of commercial and central banks (Tables 1 and 2). In 1900 gold was still used in active circulation, though it represented less than 15 percent of the world money supply, but was the main component of international reserves; in 1913 official gold amounted to more than 85 percent of total official reserves (Table 3).

The era of the gold standard ended with the outbreak of World War I. The financing of the war first and of postwar reconstruction later forced sharp and inflationary increases in the monetary liabilities of national banking systems, while gold production expanded at a much lower rate than previously. The ratio of gold reserves to money supplies fell drastically and convertibility was suspended in a large part of the world.

2.2. In the early twenties attempts were made to reconstruct a viable financial system after the turmoil created by the war. In 1922 the Genoa Conference recommended the adoption of a gold exchange standard: central banks were urged to integrate their gold with foreign exchange reserves in order to economize on the use of the existing stock and the future output of gold. An attempt was thus made to create a monetary order which would combine a pure fixed exchange rate mechanism of balance of payments adjustment, modeled on the classical gold standard, with a new mixed commodity-currency standard to cope with the shortage of gold.

From the point of view of economizing on gold the system was quite successful for a few years.[1] During the depression of the thirties, however,

[1] The amount of official gold increased by some 270 million ounces from 1913 to 1928; however its share in total reserves declined by 10 percentage points reflecting the increase of foreign exchange reserves, which at the end of 1928 amounted to 3.2 billon dollars (0.7 billion in 1913).

Table 1 - *Gold 1800-1981* (estimated amounts in million ounces)

	Current flows			*End of period stocks*			(B) / (A) = *Monetary Stocks/ Total Stocks*
		of which			*of which*		
	Production	*Non Monetary Uses*	*Monetary Uses*	*Total* (1) (A)	*Non Monetary*	*Monetary* (2) (B)	
1801-1848	31.4	16.9	14.5	146.2	93.0	53.2	36.4
1849-1913	588.7	246.1	342.6	734.9	339.1	395.8	53.9
1914-1928	270.8	117.3	153.5	1,005.7	456.4	549.3	54.6
1929-1933	102.4	48.2	54.2	1,108.1	504.6	603.5	54.5
1934-1937	114.4	—57.7	172.1	1,222.5	446.9	775.6	63.4
1938-1949	309.5	49.2	260.3	1,532.0	496.1	1,035.9	67.6
1950-1959	314.4	156.7	157.7	1,846.4	652.8	1,193.6	64.6
1960-1969	451.0	425.7	25.3	2,297.4	1,078.5	1,218.9	53.1
1970-1980	442.0	461.0	—19.0	2,739.4	1,539.5	1,199.9	43.8
1981(*)	37.3	33.8	3.5	2,776.7	1,573.3	1,203.4	43.3

(1) Includes sales by Communist countries. In the period 1492-1600 estimated production of gold amounted to 24 million ounces; in 1601-1700, 29 million and in 1701-1800, 61 million ounces.

(2) Monetary gold stocks includes gold in active circulation, which at the end of 1913 was estimated at 160 million ounces.

(*) Estimates.

Sources: BIS, IMF and R. Triffin: *The Evolution of the International Monetary System: Historical Reappraisal and Future Perspectives*, Princeton Studies in International Finance, n. 12, 1964.

Table 2 - *Estimated sources and uses of gold 1929-1981* (million ounces)

	1929-1980	*1929-1933*	*1934-1937*	*1938-1949*	*1950-1959*	*1960-1969*	*1970-1980*	*1981(*)*
Western production	1,548.1	102.4	105.0	313.5	280.0	386.5	360.7	30.1
South Africa	946.7	54.6	35.8	151.3	145.4	280.8	278.8	21.1
Estimated sales by Communist countries	185.6	—	9.4	—4.0	34.4	64.5	81.3	7.2
Total	1,733.7	102.4	114.4	309.5	314.4	451.0	442.0	37.3
Change in official gold (monetary uses)	650.6	54.2	172.1	260.3	157.7	25.3	—19.0	3.5
Non monetary absorption	1,083.1	48.2	—57.7	49.2	156.7	425.7	461.0	33.8

(*) Estimates.

Sources: BIS, IMF, R. Triffin: *The Evolution of the International Monetary System: Historical Reappraisal and Future Perspectives*, Princeton Studies in International Finance, n. 12, 1964 and F. Hirsch, *Influences on Gold Production*, IMF, Staff Papers, vol. XV, n. 3, November 1968.

Table 3 - *Gold in international official reserves 1913-1981* (1) (in percent, end of period)

	1913	*1928*	*1937*	*1950*	*1960*	*1970*	*1980*	*1981(*)*
Official gold as percent of international official reserves (2)	85.4	75.7	91.4	68.8	63.6	41.3	62.6	50.0
Distribution of official gold among main groups of countries								
— Industrial countries	—	—	—	90.9	90.8	86.5	84.6	84.6
— Oil-exporting countries	—	—	—	2.2	1.9	3.2	3.9	4.0
— Non oil developing countries	—	—	—	6.9	7.3	10.3	11.5	11.4
(Total)				(100.0)	(100.0)	(100.0)	(100.0)	(100.0)

(1) Gold valued at $20.67 an ounce for 1913 and 1928, at $35 for 1937 and at market prices for the period 1950-1980.

(2) International official reserves = gold + foreign exchange + SDR's + reserve position in the Fund.

(*) Estimates.

Sources: IMF and R. Triffin: *The Evolution of the International Monetary System: Historical Reappraisal and Future Perspectives*, Princeton Studies in International Finance, n. 12, 1964.

competitive devaluations of national currencies in relation to gold had the effect of endowing gold with capital value gains and a resultant positive rate of return, which tended to exceed that available on foreign currencies. Countries thus took action to protect and even increase their gold reserves, which at the end of 1937 came to as much as 91 percent of their total reserves. It must be noted that, because of the fallacy of composition, this adjustment could be achieved only at the cost of the Great Depression. The ensuing devaluations of the principal currencies vis-à-vis gold (the price of an ounce went up from $20.67 to $35 and in pound sterling from 4.24 to 8.66) can thus be described as a means of achieving a substantial (some 80 percent) writing up of what at the time could be properly defined as the world monetary base in terms of currencies.

2.3. The monetary aftermath of World War II presented a number of contrasts, but also a striking similarity with that of World War I. Once again wartime and postwar reconstruction financing brought about large increases in money supplies and a considerable decline in the ratio of international reserves to domestic monetary (paper) assets. The Bretton Woods system represented a monetary order combining an essentially unchanged gold exchange standard, supplemented by a centralized pool of gold and national currencies, with a new adjustable peg exchange rate mechanism. The negotiators at Bretton Woods did not think it necessary to alter in any fundamental way the mixed commodity-currency standard that had been inherited from the interwar years. The problem in the twenties, they felt, had not been the gold exchange standard itself but the division of responsibilities. The nucleus of the gold exchange system resided in more than one country and this proved a source of weakness. It was felt, however, that with adequate cooperation between the centre countries this need not create fundamental difficulties. In the Bretton Woods system, the IMF was to provide the necessary machinery for multilateral cooperation.

During the fifties the world developed a temporarily very successful gold-dollar exchange standard. The United States had emerged from World War II with an overwhelmingly dominant economy in terms of productive capacity and national wealth, including reserves of monetary gold — by 1940 US official gold had climbed to 70 percent of all official gold (Table 4). In the eyes of the rest of the world dollar holdings were more desirable than gold, since they were readily exchangeable into the metal and brought a positive interest to their holders.[2] This evolution is reflected in the fact that in 1960 official gold had fallen to 64 percent of international official reserves, while the share of the dollar in foreign exchange reserves had risen to more than 50 percent, i.e. nearly twice as much as in 1949, at the expense of the pound

---

[2] This positive rate was, however, very low compared with that obtaining on other currencies, as will be argued in the following section.

Table 4 - *Main holders of official gold 1937-1980* (million ounces, end of period)

	1937	*1940*	*1945*	*1950*	*1955*	*1960*	*1965*	*1970*	*1975*	*1980* (¹)
All countries	722.6	894.3	965.7	949.5	999.9	1,083.3	1,193.6	1,057.9	1,017.8	1,035.0
United States	365.4	629.8	573.8	652.0	621.5	508.7	401.9	316.3	274.7	264.3
Germany	—	—	—	—	26.3	84.9	126.0	113.7	117.6	114.2
France	78.5	64.9	32.9	18.9	26.9	46.9	134.5	100.9	100.9	102.3
Italy	6.0	3.4	0.7	7.3	10.1	63.0	68.7	82.5	82.5	83.3
Switzerland	18.6	14.3	31.5	42.0	45.6	62.4	86.9	78.0	83.2	83.3
United Kingdom	116.7	8.4	56.6	81.8	57.5	80.0	64.7	38.5	21.0	23.5
Argentina	12.7	11.9	34.2	2.4	6.0	3.0	1.9	4.0	4.0	4.4
India	7.8	7.8	7.8	7.1	7.1	7.1	8.0	7.0	7.0	8.6
Percentage share of:										
United States	50.6	70.4	59.4	68.7	62.2	47.0	33.7	29.9	27.0	25.5
Italy	0.83	0.38	0.07	0.77	1.01	5.82	5.76	7.80	8.11	8.04

(¹) Including gold swapped with the EMCF as counterpart of ECU creation.

Source: IMF.

Table 5 - *Official foreign exchange reserves by currency and holders* (end of period, billion dollars and % shares)

	1949	*%*	*1962*	*%*	*1970*	*%*	*1975*	*%*	*1980*	*%*
Currencies										
Dollar	3.2	27.4	12.9	57.3	34.3	74.6	126.7	78.8	243.0	74.8
Official dollar claims on US					23.8	51.8	80.7	50.2	158.3	48.7
Eurodollars and others					10.5	22.8	46.0	28.6	84.7	26.1
Pound Sterling	6.4	54.7	6.2	27.6	6.2	13.5	8.0	5.0	9.1	2.8
Official sterling claims on UK					5.7	12.4	7.5	4.7	6.9	2.1
Eurosterling and others					0.5	1.1	0.5	0.3	2.2	0.7
Deutsche Mark	—	—	—	—	1.3	2.8	10.5	6.5	42.0	12.9
Official DM claims on Germany							2.9	1.8	15.0	4.6
Euromarks and others							7.6	4.7	27.0	8.3
Other currencies	2.1	17.9	3.4	15.1	4.2	9.1	15.6	9.7	30.9	9.5
Total	11.7	100.0	22.5	100.0	46.0	100.0	160.8	100.0	325.0	100.0
Memorandum item: Official gold at market prices	32.6		41.2		39.5		142.4		610.5	
Holders										
Industrial countries	5.5	46.7	15.6	69.5	31.0	67.4	81.0	50.4	160.6	49.4
Oil-exporting countries	0.5	4.5	1.3	5.7	4.0	8.7	50.0	31.1	85.4	26.3
Non oil developing countries	5.7	48.8	5.6	24.8	11.0	23.9	29.8	18.5	79.0	24.3

Source: BIS, IMF, OECD and estimates based on R. Triffin: *The Evolution of the International Monetary System: Historical Reappraisal and Future Perspectives*, Princeton Studies in International Finance, n. 12, 1964.

sterling (Table 5).

2.4. In the early sixties the value of US short term external obligations came to exceed the value of US gold holdings, which in turn had gradually fallen back to less than 50 percent of total official gold. The first symptoms of serious difficulty on the Bretton Woods system appeared. They took the form of speculative private sales of dollars and some official demand for gold, in expectation of a dollar devaluation (in the period 1960-65 the United States lost 107 million ounces of its official gold; in the same period France and Germany increased their gold holdings by 88 and 42 million ounces respectively). In an effort to contain speculative pressures in 1961 a "gold pool" was created among the major financial powers to stabilize the price of gold in private markets. In 1968 this gold pool was replaced by a two tier gold price system, one price on private markets determined by supply and demand and another price for central banks to remain at the previous fixed level of $35 per ounce.

In 1971 the Bretton Woods system collapsed. During the last decade the official gold has been virtually "frozen"; none of the newly mined gold has gone into the monetary gold stocks.[3] The official price of gold (35 SDR per ounce since 1971) was definitively abolished in 1978 when the Second Amendment of the Fund came into force.

2.5. At the beginning of the eighties the role and the prospects of gold in the international monetary scene must be analyzed in the context of a multicurrency reserve system. A revival of the gold standard is neither desirable nor practically feasible. It seems likely, however, that gold — although imperfectly liquid — will remain an important component of official reserves.

A main problem in this regard is how to mobilize it, in case of need, without affecting the private market. An example of partial mobilization is borrowing by using gold as collateral, as Italy did from Germany in 1974. Another way is that adopted within the EMS, which provides for ECU creation vis-à-vis gold swaps with the EMCF. It should be pointed out, however, that the technique adopted for the contributions of gold and dollars to the EMCF entails effects of a predominantly accounting nature; in fact the EMCF does not have full disposal of the international reserve assets transferred to it, while the corresponding ECU-denominated reserves are not created on a stable or definitive basis.

---

[3] Available statistics do not indicate any increase in the volume of official gold stocks. It can, however, be assumed that some new gold has been acquired by the OPEC countries. Moreover, in the recent phase of declining prices some small central banks and monetary authorities have been net gold buyers. In fact the official sector which in 1980 became — for the first time since 1972 — a net absorber for gold rather than a net supplier has continued during the past year to take significant (3.5 million ounces) quantities of metal from the market.

3. Gold and Paper Assets: a Risk-Return Assessment in the Postwar Perspective

The emergence of a multicurrency reserve system occurred gradually in the seventies and was intimately linked, on the one hand, to the recourse to floating, but heavily managed, exchange rates and, on the other, to the permanent imbalances in the network of current account imbalances following the two oil shocks. This led both to structural surpluses, and hence to the growing importance of yield consideration in the deployment of reserves, and to persistent deficits — which were financed largely through compensatory borrowing in foreign exchange markets. Net liability positions were thus built up, and the choice of the currency in which borrowing was to take place became a relevant policy option.

3.1. In this section an attempt is made to evaluate the performance of gold as an official international reserve asset in the postwar period and to shed some light on the working of the international monetary system, by simulating the results, in terms of yield, variance and covariance of nominal and real returns, which would have been obtained by investing alternatively in the principal currencies (dollar, pound sterling, yen, Deutsche Mark, French franc), in the SDR basket (according to its present definition), and in gold itself.

Nominal yields are all expressed in terms of dollars, by taking into account the own yield, as measured by short term money market rates, and the exchange rate change vis-à-vis the dollar. The yield on gold is represented by the change in its market price in dollars. Yearly yields are compounded to calculate cumulative returns over the period 1949-1981. Real rates are obtained by deflating nominal yields by the rate of change in the dollar unit value of world imports.

3.2. It is often argued that gold as an investment and international reserve asset is held for different motives. In particular, two opposite views have been advanced: according to the first, gold must be considered basically as any other means of holding reserves and must therefore promise an expected return, in terms of price appreciation, comparable to currency reserves, where direct yield and exchange-rate movements must simultaneously be taken into account.

Others maintain instead that gold is essentially held with a view to reducing the overall risk of diversified portfolios: this line of argument is predicated on the assumption of a negative covariance between (all) paper assets and gold at times of severe shocks.

To some extent an explanation for these different approaches can be found in the fact that they implicitly refer to different time periods, which are characterized by changing institutional features[4], and that the behaviour

---

[4] In particular, the central banks' choice between alternative international reserve assets has

of central banks has itself changed over time, but has often been more directly influenced by considerations which are not of an immediate economic nature.

However, we would argue that, especially under present circumstances, in which a fully-fledged multicurrency standard is gradually establishing itself and is likely to stay in the foreseeable future, a unitary approach can tentatively be sought along portfolio lines, in the context of imperfect asset substitutability, following models developed notably by Branson, Dornbusch, Kouri, Solnik and Szego.

This approach requires special qualifications for the analysis of the behaviour of official monetary agencies endowed with the specific task of managing foreign international claims, since their a priori preferential reasons for holding different assets can only be found in either institutional considerations or in economic factors such as the currency composition of foreign trade and/or foreign debt.

Once these preferences are specified, optimal multi-asset portfolio combinations can be obtained through a utility maximization procedure from a vector of expected returns and a matrix of expected variances and covariances around these returns. As can be shown, the optimal portfolio turns out to be a linear combination of a minimum-variance (hedging) portfolio, that depends only on variances, and of a speculative zero-net-worth portfolio in which a preferred risk-return combination is obtained by borrowing in some currencies and lending in others.

For the reasons we have already outlined it is clearly not possible to assume that central banks could have followed a multi-asset portfolio approach over the period 1949-1981. And indeed we have anyhow strong reservations about the common practice of treating ex-post mean returns and covariances as if they represent structural parameters from an ex-ante point of view.

The exercise undertaken here of calculating nominal and real yields must therefore be seen purely as a backward simulation, and no attempt is made to use the results obtained in order to determine "optimal portfolios". We feel, however, that the empirical findings are in themselves of some interest and can help to explain the development of the international monetary system in the postwar period and point to some implications for its future working.

---

been clearly constrained by institutional factors. As is known, even the choice between gold and dollar was by no means a free one during the sixties. In the seventies, the holding of currencies other than the dollar was limited to the Euromarket, and G-10 central banks voluntarily agreed to contain their diversification. It is only in the past two years that the German and Japanese monetary authorities have accepted, although reluctantly and with significant limitations, an official reserve role for their respective currencies.

3.3. A first, somewhat surprising result is that over the 32 years from end-1949 to end-1981 the yen emerges as by far the highest yield asset (Figure 1). In nominal terms, an initial amount invested in the Japanese currency would have grown by 16.5 times in terms of US dollars at the end of 1981: this is to a significant extent due to the relatively high domestic nominal yields which obtained in Japan between 1950 and 1970.

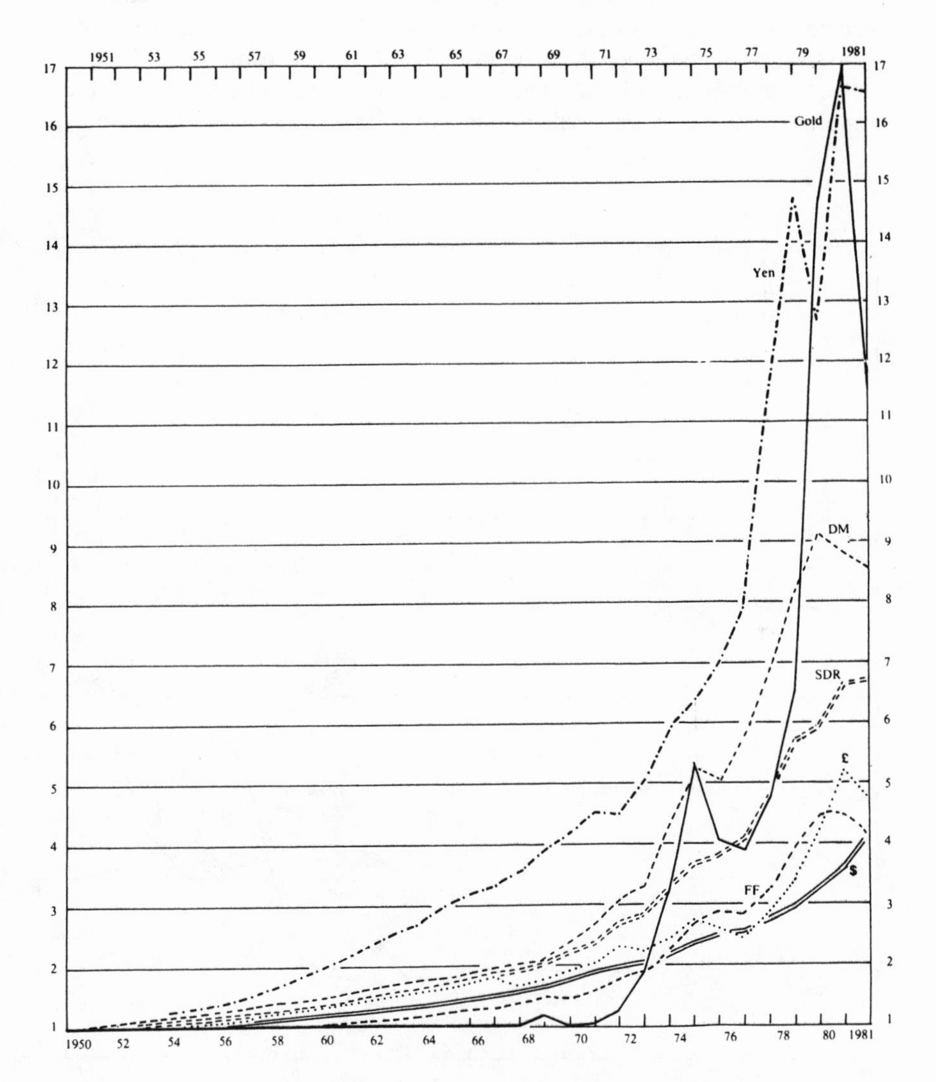

Fig. 1 Returns in dollar terms of an investment in gold, SDR, $, Yen, DM, FF and £ (31-12-1949 = 100.0).

The appreciation of the gold price in the past ten years led the yellow metal temporarily into top position between 1979 and 1980, but at the quotations of end-1981 gold has fallen back into second place, although still with an over tenfold increase in the initial investment.

The DM appears as the third-best investment in terms of straight yield. It is remarkable, however, that the overall cumulative return is only one-half that recorded by the Japanese yen. The SDR — defined on the basis of today's five-currency basket — shows a regular growth, as could be expected, but its overall performance is pulled back by the relatively poor performance of the pound, the dollar and the French franc.

The latter currencies in fact show somewhat similar end-results, with total cumulative returns of between four and five times the initial investment, i.e.

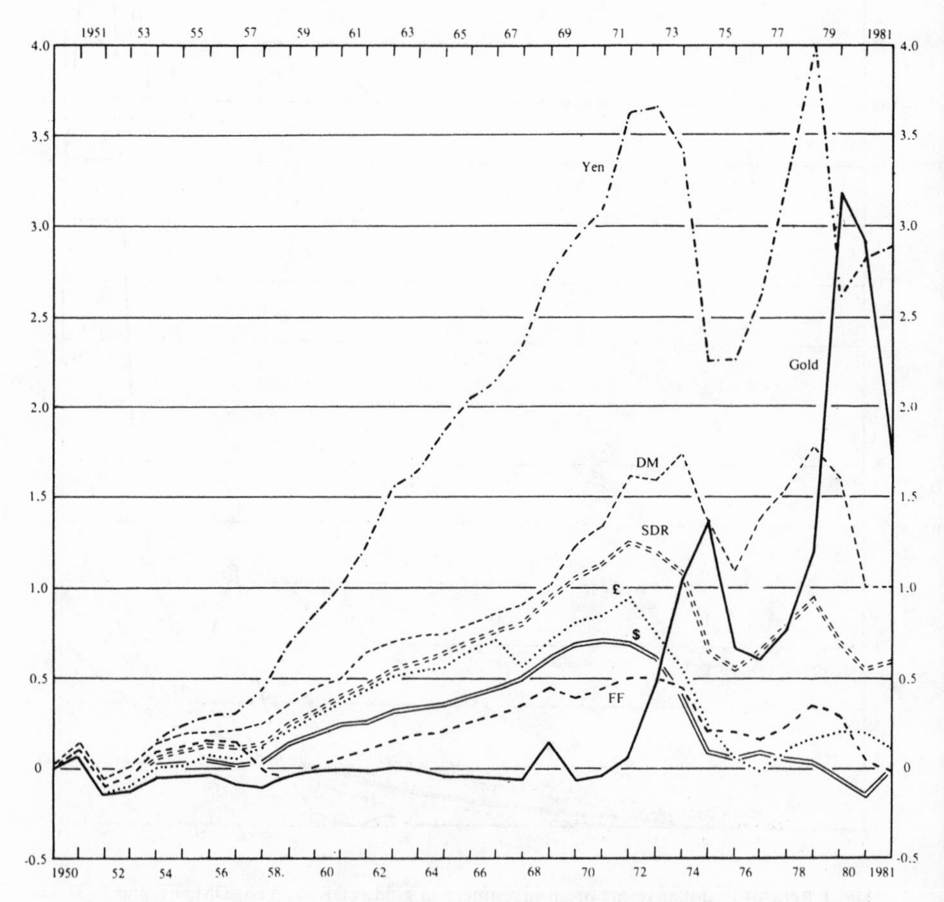

Fig. 2 Real returns of an investement in gold, SDR, $, Yen, DM, FF, and £ (31-12-1949 = 0.0)

approximately one-fourth of those on the yen. It is also noteworthy that the dollar posts an exceptional gain in 1981 as a result of the interaction between high own yield and exchange rate appreciation.

These nominal results have also been translated into real terms by subtracting the yearly percentage change in the unit value index of world imports (Figure 2). In order to assess these real yields it appears useful to subdivide the 32 years under consideration into three subperiods. From 1950 to 1970 the unit value of internationally traded commodities showed relatively small fluctuations around a flat trend. During this first period all paper assets thus showed a positive return. In particular, the yield on dollars, although far below those on yen and DM assets, was still sufficient to make the American currency superior to gold, whose price remained fixed in dollar terms, and thus showed no trend difference with respect to all other internationally traded goods.

Between 1971 and 1980, the sharp rise in world trade prices was not accompanied by a similar adjustment in dollar yields. The return on dollar assets thus became sharply negative and by end-1980 the losses had more than wiped out all previous gains: the cumulative yield over the period 1950-80 had thus come to -15 percent. As a counterpart to this very poor performance of the dollar, gold showed a very sharp appreciation also in real terms.

In 1981 the shift in the attitude of the American administration to domestic monetary policy, and the abatement in world trade prices[5], made for an unprecendented rebound of the "real" performance of the dollar.

Over the whole period 1950-1981 the cumulative real yield on dollars thus comes back exactly to zero. At the other extreme, the yen shows an overall real return of nearly 300 percent, while gold and the DM post returns of 170 percent and 100 percent respectively.

If, as has been argued, a generalized portfolio approach is in principle appropriate to ascertain the desirable composition of multi-asset international reserve portfolios, the straight-yield considerations developed so far must be supplemented by analysis of the variance and covariance of returns.

As can be seen from Tables 6a and 6b, these results too point to the desirable properties of the yen. Not only does it show the lowest variance in nominal terms, but it also exhibits a negative covariance with gold, which, as we have just indicated, has the second highest yield. While the extreme hedging hypothesis on gold recalled above refers to very severe shocks only, it must, however, be recorded that the data indicate a relatively high positive association between the precious metal and the dollar. Finally, we observe

[5] The unit value of world imports showed last year an estimated 2% fall — the first absolute decline recorded since 1968.

Table 6a - *Means, variances and covariances of nominal returns of an investment in five major industrial countries, in SDR and in gold: (A) 1950-1981, (B) 1950-1970, (C) 1971-1980.*
(percentage points)

	(A) 1950-1981						
Statistic	*United Kingdom*	*Germany*	*France*	*Japan*	*United States*	*SDR*	*Gold*
Geometric mean	4.96	6.94	4.54	9.16	4.62	6.12	7.94
Variance or covariance							
United Kingdom	13.2	—	—	—	—	—	—
Germany	4.8	5.7	—	—	—	—	—
France	9.3	5.0	8.8	—	—	—	—
Japan	3.0	0.6	2.1	5.5	—	—	—
United States	9.9	4.7	7.9	1.6	9.3	—	—
SDR	8.4	4.4	6.9	2.1	7.3	6.2	—
Gold	52.0	27.5	36.5	—9.2	29.2	27.8	930.4

	(B) 1950-1970						
Statistic	*United Kingdom*	*Germany*	*France*	*Japan*	*United States*	*SDR*	*Gold*
Geometric mean	3.52	4.73	2.31	7.50	3.16	4.21	0.35
Variance or covariance							
United Kingdom	4.0	—	—	—	—	—	—
Germany	0.5	2.2	—	—	—	—	—
France	2.4	1.0	2.4	—	—	—	—
Japan	2.0	—0.8	1.0	4.8	—	—	—
United States	2.8	0.9	2.2	0.9	2.6	—	—
SDR	2.3	0.9	1.9	1.2	2.0	1.7	—
Gold	0.6	—0.8	0.1	0.3	0.2	0.1	32.8

Table 6a (continued)

Statistic	(C) 1971-1980						
	United Kingdom	*Germany*	*France*	*Japan*	*United States*	*SDR*	*Gold*
Geometric mean	9.60	12.80	10.88	13.76	6.80	10.06	31.76
Variance or covariance							
United Kingdom	11.5	—	—	—	—	—	—
Germany	3.1	7.9	—	—	—	—	—
France	5.5	3.8	5.7	—	—	—	—
Japan	4.6	3.3	3.9	8.1	—	—	—
United States	8.1	2.7	3.8	2.9	6.1	—	—
SDR	6.8	4.0	4.3	4.0	5.0	4.8	—
Gold	44.9	53.5	32.2	—48.1	44.3	32.6	1876.6

Table 6b - *Means, variances and covariances of real returns of an investment in five major industrial countries, in SDR and in gold: (A) 1950-1981, (B) 1950-1970, (C) 1971-1980.*
(percentage points)

Statistic	(A) 1950-1981						
	United Kingdom	*Germany*	*France*	*Japan*	*United States*	*SDR*	*Gold*
Geometric mean	0.33	2.21	—0.06	4.34	0.0	1.44	3.16
Variance or covariance							
United Kingdom	9.0	—	—	—	—	—	—
Germany	2.6	3.7	—	—	—	—	—
France	6.2	3.0	5.9	—	—	—	—
Japan	1.2	—0.6	0.5	4.7	—	—	—
United States	7.1	3.3	5.8	0.3	7.5	—	—
SDR	5.6	2.7	4.7	0.9	5.5	4.3	—
Gold	25.9	9.2	13.5	—19.0	10.7	8.9	665.7

Table 6b (continued)

Statistic	(B) 1950-1970						
	United Kingdom	*Germany*	*France*	*Japan*	*United States*	*SDR*	*Gold*
Geometric mean	2.95	4.16	1.77	6.93	2.61	3.66	—0.2
Variance or covariance							
United Kingdom	3.8	—	—	—	—	—	—
Germany	0.4	1.8	—	—	—	—	—
France	2.3	0.9	2.4	—	—	—	—
Japan	2.0	—0.8	1.1	5.1	—	—	—
United States	2.7	0.8	2.0	0.8	2.5	—	—
SDR	2.3	0.7	1.8	1.3	1.9	1.6	—
Gold	0.9	—0.8	0.3	0.6	0.5	0.3	32.9

Statistic	(C) 1971-1980						
	United Kingdom	*Germany*	*France*	*Japan*	*United States*	*SDR*	*Gold*
Geometric mean	—4.21	—1.50	—3.10	—0.61	—6.70	—3.93	15.07
Variance or covariance							
United Kingdom	6.5	—	—	—	—	—	—
Germany	0.3	3.9	—	—	—	—	—
France	2.2	0.7	2.1	—	—	—	—
Japan	1.7	0.7	1.0	4.5	—	—	—
United States	4.3	0.6	1.3	0.8	3.3	—	—
SDR	3.2	1.2	1.4	1.4	2.3	2.0	—
Gold	14.6	24.7	4.5	—52.6	21.0	9.1	1305.0

that among currencies, the pound is the one which is characterized by the highest variance, and the second highest is recorded by the dollar itself.

4. Conclusions: a Multipolar International Monetary System and the Possible Role of Gold.

Economic as well as political considerations suggest that the present system of multi-asset relationship is likely to constitute the international monetary scenario in the eighties. A multipolar approach need not represent a "bad" solution; it is, however, clear that the viability of the system depends on the achievement of a "workable" degree of competition between the major alternative reserve assets. From this point of view, the relatively poor performance — in nominal and especially in real terms — of the dollar in the seventies documented in this study was clearly not sustainable. Highly negative real rates of return on dollar assets no doubt contributed to the inflation which has characterized the past decade and to the turmoil in exchange rate relationships. The new approach to the control of monetary aggregates pursued by the FED since 1980 can thus be seen as a way of implementing a necessary, more rigorous line in monetary policy.

It is indeed to be hoped that, barring shocks from the fiscal side, once the present somewhat dogmatic policy has achieved a reasonable degree of success in dampening domestic inflation, more attention will be devoted to the need to coordinate economic and especially monetary policies between major reserve-currency countries.

In this multipolar world — where the SDR and the ECU could also play a useful role if the IMF and the EEC were able to provide a well-defined high-powered core to their establishment in the private markets — it would be unwise to relegate gold to a limbo position. Its main property of representing a final commodity reserve asset which has no counterpart liability position adds a significant degree of choice to international investors in the selection of an overall diversified portfolio.

While it would be unwise to attempt any effort to fix the price of gold in terms of currencies, we would not rule out the possibility that under specific circumstances a two-way relationship could be re-established between the monetary and the non-monetary markets for gold. In the short run, however, it would appear desirable to concentrate efforts on workable mechanisms for multilateral mobilization of gold between official monetary agencies. As has already been indicated, a revision of the present ECU creation system might be a useful step in this direction. Similar developments could indeed take place in the broader IMF framework, also if the idea of a Fund fully based on the SDR goes ahead.

More generally, in order to pursue the objectives of (i) securing a reasonable degree of stability in the functioning of the international monetary system, and then (ii) proceeding with fully-fledged plans of reform based on the re-establishment of some forms of mandatory asset settlement in the context of a return to fixed par values — albeit with appropriate flexibility — what is basically required, in the short term, is continuation of the present policies of rigorous control of domestic monetary base creation in the major reserve-currency countries and fostering of structural adjustment in the economic systems to changes in relative prices.

CHAPTER 23

OBSTACLES TO A RETURN TO THE GOLD STANDARD

HENRY C. WALLICH

I feel it is a sort of speaking *ex cathedra* here, but what I am going to say will nevertheless be my own views on the matter. I have enjoyed the conversation here today. Mrs. Schwartz and I and Mr. Coyne, who I think is here in the audience, have been exposed to a great deal of conversation about gold in the Commission to which I have the honour to belong, and while I would not agree with everything that I have heard here, I must say that the consistency of viewpoints has not compared unfavourably with what we have heard in Washington.

As Mrs. Schwartz has said, there is a deep-seated concern about inflation. Some people think that one can come to grips with inflation via a return to gold. I think the people in the States who are in favour of some kind of return to gold are generally reliable anti-inflationists. They want something that is highly desirable; that we all want. It is a question of technique, of the means, rather than of the objectives that we are concerned about.

In a very short time you will be able to see the report of this Commission that Mrs. Schwartz will write and that will be available hopefully by March 31. I think you will find represented quite a range of views. It may turn out that we may not have a single report, but dissenting opinions or at least footnotes.

One of the members of that group said that until we have rendered that report the United States government really does not have a gold policy. Well, I do not know if that is true or not, but I am fairly confident that the situation will not change fundamentally after this report will have been rendered. It will be a report; it will have to be acted upon by the Congress if the Congress wants to. Once that is done we will indeed have a gold policy. But I would not say that at the present time from past practices, we could not say that there are the elements of a gold policy, too, although indeed there have been quite a few changes: for instance, the move from option auctioning off gold to the more recent attitude not to do that.

1. A Contemporary View of the Gold Standard

Let me try to express my views about the approach to a gold standard and gold policy and then conclude by saying something about what, I think, the United States should do.

I do not share the Keynesian belief that the gold standard is a barbarous relic and nothing else. This view, which many economists who were educated during the 1930's and 1940's became imbued with, was caused by the calamities of the 1930's when the gold standard meant depression, unemployment, currency disorder, exchange control. These were the results of last-ditch efforts made in Germany, Britain and many other countries to hold the gold standard. As a Chilean President once said, he was going to defend the gold standard "hasta con las armas", and he did but was turned out by a revolution. The gold standard got a very black eye at that time.

But the gold standard functioned very well during the 19th century. To me, at least, that was a kind of golden age of very rapid growth, currency stability and peace, relatively speaking. Mrs. Schwartz has pointed out this stability was indeed of a relative kind: we had price stability in the long term, but there were price-level fluctuations in the short term and, of course, long periods of high unemployment. It was no bed of roses.

In short, the attitude towards a gold standard that prevails now among a great majority of economists in the United States is one of somewhat excessive arrogance. Given the record of our handling of our affairs, I think there is probably something to be said for an alternative system. I have not come to be a supporter of that system, but I think it is worth discussing. As Anna Schwartz said, if we do not improve our handling of our affairs, then we may find that American politics becomes increasingly influenced by a gold debate just as it was influenced by a silver debate at the end of the 19th century (and incidentally the "silverites," who were the inflationists of those days, were defeated by the hard-money people).

2. Three Reasons for Opposing the Gold Standard

Let me examine some of the reasons why I am skeptical about a renewal of the gold standard. One is that I doubt very much that we could get there. One good reason is enough so I would not have to go much further than that, but I do want to go further. I want to argue that even if we could get there it would not work well, and even if it would work well for a time it would not work well in the long run. So, these are all cumulative reasons arriving at the same conclusion.

2.1. *The Reentry Problem*

Starting with the most immediate problem, we have what has become known as the re-entry problem, that is, how do you get back to a fixed price of gold. Most plans I have seen involve a fixed price of gold and some convertibility. Whether at the official level only as in the Bretton Woods system or convertibility for the public as well is a secondary matter. It is conceivable to have a kind of gold standard as Mrs. Schwartz described it in which there is no convertibility. All you have would be a fixed relation between the gold stock and the money supply. There would be a rule for writing up the value of the gold stock. This would mean a money supply rising at the same rate. That is a money-supply rule, not a gold standard. Any meaningful gold standard would have to be one in which at least the dollar was convertible into gold both ways: someone bringing gold gets dollars, someone bringing dollars gets gold. This might apply only to central banks, perhaps also to the public. Whether or not other countries would follow suit was not discussed a great deal in the Gold Commission. Historically, of course, the gold standard was primarily an international instrument. It was a means of maintaining a stable exchange rate. To that objective all other goals were subordinated: price stability, full employment, and growth. A price rule was the only rule.

To leave out the international aspect, and thereby leave out the fixity of the exchange rate, is really to talk about only one-half of the gold standard, but one can, of course, talk about that one-half.

Let me examine the problems that would present themselves if the United States tried to get back to a gold standard and were determined to fix a price. What price? Well, there is a price in the market. You can say that the market should decide what the price should be when a country goes back to gold. You can look at gold futures and you will see what the price of gold is expected to be 12 months from now. But one will probably fool oneself: one would observe mostly the interest cost of carrying the metal rather than the expected future gold price. At any rate, the present market price, and this is recognized by most proponents, is not a useful guide for going back to gold. What the proponents typically argue is that it would be announced that the United States was going back to gold and would fix a price some time in the future, six months or a year ahead. Then the market would evolve in the knowledge of that future event.

What will happen to the price, if such an announcement were made? I hope you realize I am speaking entirely about a hypothetical event which I do not believe for a minute will occur. I am going through this exercise in order to show some of the problems. There are two types of view in the market about the reasons for holding gold and, as Mr. Masera has stated them, they are not necessarily alternatives. They can be held simultaneously by different holders. One is that gold is an asset like any other and one has to expect the same rate of return from gold as from other assets of comparable risk which one would get from an appreciation of gold. The other view is that gold is a means of diversifying a portfolio. If there is a negative correlation between the price of gold and the price of other assets, like common stocks for instance, then one may take gold into one's portfolio even at a low or zero or even negative rate of return because of the advantage it gives one in diversification. If one reduces the risk of the entire portfolio, one can then choose other assets that have a higher probability of appreciating, knowing that one has protected oneself by this greater diversification. I do not know what the diversification advocates would do if they knew that the price of gold was going to be fixed, and I mean, of course, credibly fixed. But I do think I know what investors who hold gold for appreciation would do. If they are told that gold is not going to appreciate any longer, the logical thing is to dispose of it as soon as possible. So, if there is a substantial body of investors in the market of this persuasion, they will probably act accordingly and, following such an announcement, the price of gold will probably drift down. People who think in terms of a continuous secular appreciation of gold have nothing more to fear than a credible fixing of the price of gold.

One way of proceeding might be, then, to wait for what the market will say, half a year or a year after the initial announcement, and proceed then to a fixing. Could such a fixing be credible? There are two immediate consequences. Suppose that the price that is fixed turns out to be "high" in some sense with respect to what the market thinks. In that case we will probably have a repetition of the 1930's, when we had the "golden avalanche." Gold flooded to the United States creating excess reserves for the banks. It would have created a strong monetary expansion, had it not been for the depressed nature of the times. So, too high a price would mean that gold flows into the United States, inflates our money supply, and adds to inflation.

Another possibility is that the price is fixed at a low level. The world might then think that the United States does not have enough gold to hold that price, and might begin to pull gold away immediately. The United States would lose its gold stock and would have to give up the experiment unless it took very drastic tightening actions through monetary policy.

The choice, therefore, of an initial price is a very critical one. It may lead to a great inflow or it may lead to a great outflow. Either way there would be a disturbance of the money supply.

2.2. *Post-Reentry Difficulties*

But unless we were able to maintain a stable price level, and perhaps a continually declining price level, we would never be setting the "right" price permanently. Even by setting the "right" price at the time we would not be out of the woods. Price levels around the world cannot be expected to be stable, even if the price level in the United States were. Neither can interest rates. Instability of either would cause gold to flow in or flow out. A rigid tie of the money supply to the gold stock would then impose possibly undesirable fluctuations in the money supply or else require a monetary policy aimed at nothing but maintaining an adequate gold stock and thereby money supply, which might be destabilizing in other respects. If other countries were on the gold standard, they would encounter similar problems at the same time. If other countries are not, their exchange rates might adjust, but this might still not allow smooth sailing for the US economy. In other words, the gold standard, rather than to create stability, might become a source of instability.

2.3. *Long Run*

In the long run, finally, the economy will grow and would have a growing demand for money and therefore for gold. Gold production, as you know, is inelastic with respect to price. It may be perversely elastic because they dig deeper and use less rich ores and therefore produce less as the price goes up. In the long run we would reconstitute the problem that the gold standard presented at the time of the League of Nation's report when they found that gold was not growing fast enough and recommended a gold exchange standard and measures to withdraw gold from circulation. We would be facing the same problem that we saw in the 1950's and 1960's when we were concerned about inadequate liquidity. Robert Triffin then was examining the problem that the dollar could not indefinitely take the place of an inadequate growth of gold. The IMF also did a study of the evolution of liquidity and could not really make it credible that liquidity from gold was increasing adequately. In other words, in the long run we would discover that the very stability of the supply of gold which is so much stressed by the gold advocates would trip up the system. There is not enough "give" in it.

From a short run point of view because of the re-entry problem, from a medium term point of view because of possibly undesirable behaviour of a gold-based money supply, and from a long run point of view because of an inadequate gold supply, I see very great obstacles to an adoption of a gold standard.

Let me turn then to some of the by-products of this debate and examine particularly a proposal to issue a gold-backed bond. Would interest rates come down? It is not generally said specifically whether these are interest rates on all securities or only on this paricular gold-backed bond. But the answer is very easy for a market technician. Any market technician can construct a plan for a gold-backed bond that would indeed have a very low interest rate. It would be a convertible bond, have a "gold kicker" in the way other bonds sometimes have an "equity kicker". In the extreme, given that there are people who are willing to hold gold without receiving interest, I do not see why they should receive any interest on a gold-backed bond if that bond is convertible into gold after a fixed period of time. They would be saving the insurance, and they would be saving the storage cost. If they are willing to hold gold they should be willing to hold what would be really a warehouse certificate for gold.

If you feel you want to pay a modest interest rate on this bond, you could make it convertible a little above the present gold price. Make it convertible at $400 and there will be some interest rate that will have to be paid to get people to accept it. Make it convertible at $10,000, and few will buy it unless you pay exactly the same interest rate that has to be paid on other securities. It is simply a question of the conversion price that would determine the interest rate on such a bond. That interest rate, whatever it is, would have very little to do with the rest of interest rates in the economy, either the government's rates on unbacked bonds or corporates or other debts.

The gold-backed bond, in other words, is only a misunderstanding, and it does not in any way lead us out of our problem of high interest rates.

To conclude, let me say what I think would be a desirable set of policies for gold. I should add, of course, that my mind is open and that if Mrs. Schwartz, or Mr. Herbert Coyne, or any other member of the Commission persuades me differently I will change my policy recommendation, but, as of now, this is what I think we ought to do.

I think we ought to do absolutely nothing with our gold but sit on it. We should not sell any of it; we should not buy more; we should not revalue; and we should not operate with it in the market. Here are the reasons. You never can tell what the future may bring. It seems very unlikely to me that the world will go back to the gold standard, and it also seems unlikely to me that there will ever be the kind of war that would make a war chest of that kind at all helpful. But, nevertheless, we have this war chest and we have this reserve in case something unexpected should happen to the world. Other countries could go back to gold. Conceivably, all countries might mismanage their affairs so horribly that the forces Mrs. Schwartz described earlier may gain strength and finally force a resumption of the gold standard. I do not believe that any of this will happen, but I think it is unwise to dispose of a reserve that one happens to have.

I would not, therefore, recommend selling gold off gradually, even though it is not an income-producing asset. Some people say that if we are not going to use it, why not sell it. I think that would be a mistake. We can mint coins and then sell medallions. That is not an unreasonable demand on our resources. But I would not replace the gold that had been disposed of through coin and medallions by new purchases. I would not write up the gold to market price. I can only see mischief coming from that. As soon as you show a very large profit, there will be somebody who will find a way to use it. Whether that use is a modest one like buying back the Federal Reserve's portfolio of government securities and putting the Federal Reserve under the budget because it had lost its income, or whether that use is the more ambitious one of saying "we have got a very large deficit, so let's sell the gold and reduce the deficit thereby," or whether that use would be just generally to reduce public debt — these are all mischievous purposes. I would be very much disturbed if we got into any of them.

If we arrive at the conclusion that nothing at all should be done, maybe we will be a little frustrated and we may feel that we have spent a great deal of time arriving at very little. I would say that it is not very little if one clarifies one's thinking and thereby avoids the considerable damage that might be the result of ill-considered new departures.

CHAPTER 24

GOLD OPERATIONS AS AN INSTRUMENT OF MONETARY POLICY

JÜRG NIEHANS

1. On the End of Bretton Woods

Like Governor Wallich, I shall consider gold from the point of view of an individual central bank. However, I shall not be concerned with the question of a possible return to the gold standard. It is true that we are now witnessing a revivalist movement in favour of such a return in the United States. It is an expression of conservative fundamentalism, an economic counterpart to "creationism" in biology. Indeed, I believe that the gold exchange standard constructed at Bretton Woods was a perfectly viable monetary system whose benefits we could still enjoy had the United States not sealed its doom by embarking on an inflationary course in the early sixties. However, the return to the gold standard is not a serious option as long as the United States has not experienced several years without inflation at, on average, satisfactory employment levels. The present policy of massive deficit spending has pushed this moment further into the indefinite future. I thus regard it as realistic to consider the consequences of central bank gold operations under present-day conditions of floating exchange rates with gold being traded at current market prices.

2. Gold from an Investment Point of View

In a recent address, Henry Wallich has recommended as the course of wisdom in the matter of central bank gold to do exactly nothing. "The most desirable gold policy", he said, "for monetary authorities is to sit on their gold as an ultimate reserve and to use it as little as possible."[1] Inasmuch as

---

[1] Henry C. Wallich, *Gold and the dollar, Remarks at the Seventh International Working Conferences* sponsored by Forex Research and The International Herald Tribune, Paris, Nov. 23, 1981.

this recommendation envisaged the liquidation or accumulation of central bank gold from an investment point of view, I shall not take issue with it. From this point of view, a decision to liquidate gold stocks would have to be based on the central bank's conviction that other forms of national wealth offer a higher social return than gold. The accumulation of gold, on the other hand, would have to be based on the conviction that the return on other assets is lower. Such convictions are difficult to justify. Day by day, the gold market sees to it that the expected rate of return on gold, implicit in expected future gold price changes, is brought in line, given the information available to the market, with the returns on other assets. To justify liquidation of accumulation, the foresight of the central bank would thus have to be superior to that of the market. This is certainly conceivable, but the experience with United States gold auctions, which turned out to have subsidized private speculators at the expense of the taxpayer, are not encouraging. I believe, therefore, that from the investment point of view, Wallich's proposition represents a realistic conclusion from recent experience.

I also believe that central banks should not try to intervene in the gold market for the purpose of stabilizing the gold price or reducing its fluctuations. Such efforts would probably fail because central banks are unlikely to be consistently better speculators than other traders. Even if successful they would serve no useful social purpose. The central banks, like private traders, should take market conditions as they find them and not try to dominate them.

3. Gold Operations of Central Banks for Short Term Monetary Policy

I suggest, however, that from the point of view of short term monetary policy, a more active use of gold operations by central banks might be indicated. If central banks, in the long run, wish to hold on to their gold stocks, they might just as well get an additional economic return by using them as a policy instrument. To the best of my knowledge this has not been tried on a large scale since the fall of 1933 when Roosevelt, before getting up in the morning, used to tell Morgenthau at what price gold should be bought on that day. His purpose was to stop the decline of commodity prices, but his understanding of money was rudimentary and his experiment was not successful. I believe, however, that we can do better today.

My argument follows in the tradition of Meade, Tinbergen, and Mundell. With frozen gold reserves, the central bank can create money essentially in two ways, namely either by buying domestic assets or by buying foreign exchange. The basic economic effects are qualitatively the same in both cases, inasmuch as interest rates decline and the currency depreciates. The quantitative mixture of these effects, however, is usually different.

Depending on the state of the economy, the central bank may thus prefer different combinations of open market operations and foreign-exchange operations. Under certain circumstances, it may even find it advantageous to switch between domestic and foreign assets without changing the money supply.[2]

Gold purchases or sales at the going market price offer a third basic technique of expanding or contracting the national money supply. Again, it can be shown that the instantaneous effects are qualitatively similar to those of open market purchases and foreign exchange purchases. Like the latter, a purchase of gold for national money results in a lower interest rate, a depreciation of the currency on the foreign exchange market and a higher domestic price of gold. It would be a serious mistake to believe that gold purchases have no effect on interest rates and exchange rates. Again, however, since gold, domestic securities and foreign exchange are imperfect substitutes, there are likely to be significant quantitative differences. One may be inclined to conjecture that, in the short run, gold operations have a comparatively strong influence on the gold price, that open market operations have a comparative advantage in influencing interest rates and that foreign exchange operations have an advantage in influencing exchange rates. On the basis of a short run asset model it can be rigorously demonstrated that under plausible and indeed almost inescapable assumptions this is indeed so (see Appendix). Depending on the business situation, the central bank may thus have good reasons to combine its open market and foreign exchange operations with appropiate gold operations.

Let me illustrate the nature of these considerations. Suppose the central bank wishes to contract the money supply, but it is concerned about pushing up interest rates and producing an overshooting of the exchange rate. Sales of gold may offer a way out of the dilemma in sûch a case. In the absence of open market qperation, domestic interest rates would still rise, but less than if domestic securities had been sold. In the absence of foreign exchange operations, the currency would still appreciate, but less than if foreign assets had been sold. In principle, it would even be possible to achieve the desired contraction of the money supply without any change in interest rates and exchange rates by combining somewhat larger gold sales with relatively small purchases of domestic securities and foreign exchange.

To give another example, suppose the central bank wishes to counteract an unwanted appreciation of the currency in the foreign exchange market, but it is reluctant to push up interest rates. In this case, it could come close to

---

[2] This is further worked out, including medium term and long term considerations, in Jürg Niehans, "Volkswirtschaftliche Wirkungen alternativer geldpolitischer Instrumente in einer kleinen offenen Volkswirtschaft", in: W. Ehrlicher and R. Richter (eds.), *Probleme der Währungspolitik, Schriften des Vereins für Socialpolitik*, Neue Folge, Bd. 120, Berlin (Duncher & Humblot) 1981, P. 55-111.

achieving its objective by combining purchases of foreign exchange with gold sales. Small domestic open market operations would probably be enough to neutralize any remaining interest effects. These examples should make the principle underlying short run gold operations sufficiently clear.

4. The Present Situation Calls for some Innovation

At the end of 1980, the central banks of the Group of Ten and Switzerland held more than 40 percent of their monetary base in the form of gold, almost 40 percent in the form of domestic assets and about 20 percent in the form of foreign exchange.[3] This means that a large part of central bank money is "backed", not by financial claims, either domestic or foreign, but by real commodities, produced by land, labour and capital like other commodities in national income and wealth statistics. With the collapse of the gold exchange standard (which I deplore), central banks were given the opportunity to buy and sell a commodity at prevailing market prices. They thus obtained a monetary policy instrument they did not possess before.

So far, this instrument has been hardly used. For the last ten years, except for the IMF and US sales, the gold component of central bank reserves, measured in ounces, has remained virtually frozen, thus providing no useful service for monetary policy. This is perhaps understandable because central bankers, like everybody else, first had to gain experience about the functioning of floating exchange rates and a free gold market. Now, a decade after the transition to floating rates, this knowledge is presumably available. I do not expect, of course, that a large volume of central bank gold transactions will emerge overnight. My suggestions about the potential policy role of gold operations under floating exchange rates are made here for the first time, they are untested and even some of the underlying theory remains to be explored. I submit, however, that the basic idea is compelling enough to merit further attention and some careful experimentation, at first perhaps on a small scale.

5. Appendix

The Analytical Basis of the Proposal: A Four-Asset Model

This appendix provides the analytical basis for the propositions presented in

---

[3] Gold (at market price) and foreign exchange (including IMF, SDR, ECU): Bank for International Settlements, *Fifty-First Annual Report*, 1981, p. 143. Domestic assets (claims on government and deposit money banks): *International Financial Statistics* (national currencies converted into dollars at year-end exchange rates).

the preceding paper. The argument is based on a four-asset model. The private sector holds domestic money (L), domestic bonds (B), gold (G), and foreign currency (F). Interest-bearing foreign assets are not considered. Capital gains or losses on bonds are disregarded for simplicity. Commodity prices are constant and normalized at l; as a consequence, real values are equal to nominal values. The banking sector is consolidated with the rest of the private sector. As a consequence, L is equivalent to the monetary base.

For each asset, the stock demand depends on the exchange rate (e), the rate of interest (i) and the price of gold (p):

$$\begin{aligned} L &= L\,(e,i,p) \\ P &= P\,(e,i,p) \\ pG &= G\,(e,i,p) \\ F &= F\,(e,i,p) \end{aligned}$$

In this short run model, the exchange rate influences asset demand through its effect on expected future capital gains. A rise in the present price of foreign currency, other things equal, tends to dampen the hope for future rises and increases the risk of future declines. This, taken in itself, stimulates the demand for domestic money, bonds and gold while the demand for foreign currency declines. An increase in the rate of interest has a positive effect on the demand for bonds while the effects on the other assets is negative. The price of gold, finally, has a negative effect on the demand for gold but a positive effect on the other demands. Denoting partial derivatives by subscripts, this can be summarized by the following assumptions:

$$\begin{array}{lll} L_e > O & L_i < O & L_p > O \\ B_e > O & B_i > O & B_p > O \\ G_e > O & G_i < O & G_p < O \\ F_e < O & F_i < O & F_p > O \end{array}$$

The demand for assets is subject to the wealth constraint

$$dL + dB + pdG + edF = O.$$

As a consequence, there are only three independent asset demand functions. It is convenient to treat the demand for foreign currency as dependent. The dependence implies that, after normalizing $p = e = 1$, the partial derivatives are related as follows:

$$\begin{aligned} L_i + B_i + G_i &= -F_i > O \\ L_e + B_e + G_e &= -F_e > O \\ L_p + B_p + G_p &= -F_p < O \end{aligned}$$

The total supplies of government bonds ($\overline{B}$), gold ($\overline{G}$) and foreign currency ($\overline{F}$) are assumed to be given; certain parts of these assets, denoted by the corresponding lower-case letters, are bought by the central bank in exchange for domestic money. This results in the central bank balance sheet

$$m = b + pg + ef$$

For each asset, demand must equal supply:

$$L = m$$
$$B = \overline{B} - b$$
$$G = \overline{G} - g$$

Taking differential and normalizing $p = e = 1$, the changes in e, i and p are seen to be related to central bank operations as follows:

$$L_e de + L_i di + L_p dp = db + dg + df$$
$$B_e de + B_i di + B_p dp = - db$$
$$G_e de + G_i di + G_p dp = - dg$$

The determinant from the coefficients on the left-hand side is denoted by Δ. It is left to the reader to verify that $\Delta < O$ is necessary for stability. An additional stability requirement is $B_e G_i - G_e B_i > O$.

The central bank can create money in three ways, namely by buying domestic bonds ($dm = db$; $dg = df = O$), by buying gold ($dm = dg$; $db = df = O$) or by buying foreign exchange ($dm = df$; $db = dg = O$). In each case, the effects on the exchange rate, the rate of interest and the price of gold can be obtained by solving the above system of three equations.

For domestic open market operations these effects are as follows:

$$\frac{de}{dp} = \frac{1}{\Delta} [G_p(B_i + L_i + G_i) - G_i(B_p + L_p + G_p)] > O$$

$$\frac{di}{dp} = \frac{1}{\Delta} [G_e(B_p + L_p) - G_p(B_e + L_e)] < O$$

$$\frac{dp}{db} = \frac{1}{\Delta} [G_i(B_e + L_e) - G_e(B_i - L_i)] > O.$$

For all three effects, the direction is clear from the signs of the partial derivatives and the stability conditions. Open market purchases thus produce a depreciation of the currency, a decline in interest rates and a rise in the domestic price of gold. The effects of foreign exchange purchases have the same signs:

$$\frac{de}{df} = \frac{1}{\Delta} (B_iG_p—G_iB_p) > O$$

$$\frac{di}{df} = \frac{1}{\Delta} (G_eB_p—B_eG_p) < O$$

$$\frac{dp}{df} = \frac{1}{\Delta} (B_eG_i—G_eB_i) > O$$

The same is true for the effects of gold purchases:

$$\frac{de}{dg} = \frac{1}{\Delta} [B_i(L_p + B_p + G_p) — B_p(L_i + B_i + G_i)] > O$$

$$\frac{di}{df} = \frac{1}{\Delta} [B_p(L_e + B_e + G_e) — B_e(L_p + B_p + G_p)] < O$$

$$\frac{dp}{dg} = \frac{1}{\Delta} [B_e(L_i + G_i) — B_i(L_e + G_e)] > O.$$

While qualitatively the same, the effects of the three policies are quantitatively different. The differences can be determined by considering the effects of shifts between central bank assets at an unchanged money supply. Consider first a purchase of gold financed by sales of foreign exchange (df = —dg; dm = db = O). Its effects are:

$$\frac{de}{dg} = \frac{1}{\Delta} (B_iL_p—L_iB_p) < O,$$

$$\frac{di}{dg} = \frac{1}{\Delta} (L_eB_p—B_eL_p) \gtrless O,$$

$$\frac{dp}{dg} = \frac{1}{\Delta} (B_eL_i—L_eB_i) > O.$$

With respect to the exchange rate, the downward pull of the foreign exchange sale is seen to dominate the upward pull of the gold purchase. Foreign exchange operations can thus be said to have a comparative advantage in influencing the exchange rate. By a similar argument, gold operations have a comparative advantage in influencing the gold price. The effect on interest rates remains unclear; neither instrument seems to have a clear comparative advantage and the effect is likely to be relatively small.

The same sort of reasoning can be applied to purchases of gold financed by bond sales ($db = —dg$; $dm = df = O$). In this case, the effects are:

$$\frac{de}{dg} = \frac{1}{\Delta} [L_p(G_i + B_i) — L_i(G_p + B_p)] \gtrless O,$$

$$\frac{di}{dg} = \frac{1}{\Delta} [L_e(G_p + B_p) — L_p(G_e + B_e)] > O,$$

$$\frac{dp}{dg} = \frac{1}{\Delta} [L_i(G_e + B_e) — L_e(G_i + B_i)] > O.$$

This means that open market operations have a relatively stronger effect on the rate of interest, while gold operations have again a comparative advantage for the gold price. In this case it is the exchange-rate effect which is uncertain and probably small.

It can be left to the reader to perform the corresponding comparison for a shift between domestic securities and foreign currency. The result simply confirms that gold operations have a comparative advantage in influencing the gold price, open market operations have a comparative advantage for domestic interest rates and foreign exchange operations for the exchange rate. In this case the effect on the gold price is uncertain and probably small.

Up to this point the model was used to determine the effects of given monetary policies. It is evident that it can also be used to determine the policy instruments required to obtain given effects. To illustrate by the example mentioned in the text, suppose the money supply should be contracted without a change in interest rates and exchange rates. The required combination of gold sales with bond and foreign exchange purchases is obtained by letting $di = de = O$ and solving for db and df in terms of dg. It turns out, not surprisingly, that the results depend only on the sensitivity of asset demands to the gold price:

$$\frac{dp}{dg} = \frac{B_p}{G_p} < O$$

$$\frac{df}{dg} = -\frac{L_p + B_p + G_p}{G_p} = \frac{F_p}{G_p} < 0.$$

The higher the sensitivity of bond demand and foreign exchange demand with respect to the gold price, the larger must be the compensating operations in the security and foreign exchange markets to neutralize effects on interest rates and exchange rates.

The model used in this paper is evidently restricted to short run asset effects. In the longer run, flow aspects would have to be included and commodity prices would become variable. As a consequence, the analysis is bound to become more complex and the results are generally different. However, since bonds, foreign exchange and gold are probably imperfect substitutes even in the long run, the economic effects of gold operations are still likely to differ from those of open market and foreign exchange operations.

CHAPTER 25

NATIONAL AND SUPRANATIONAL MONETARY AUTHORITIES' POSITION: CONCLUSIONS AND EVALUATIONS[1]

RENÉ LARRE

1. Peace between Supporters and Opponents of a Monetary Role of Gold

I have to report on what has been going on in the third study group, which was concerned with the position of national and international authorities towards gold.

To put it in some kind of perspective, I would remind the audience that, for the last 15 years or so, there has been a running discussion between the supporters and the opponents on the monetary role of gold and the two camps have been confronting their views in various arenas, including the G-10 and the IMF meetings. The veterans of those battles — there are a few still around — have kept a vivid memory of those discussions.

But, now, the big news at this meeting in Rome is that the war is over and that peace has been restored thanks to a compromise.

On the one hand, the supporters of gold have recognized that, although they might have been right in the first place in their efforts to maintain the fixed rate system and to preserve gold convertibility, it was now too late to come back to the past, so that what has been accomplished, during the last years, is final.

On the other side, the opponents have accepted the idea that the SDR system was not an answer. They have given up their effort to achieve the demonetization of gold and they are willing to consider the metal as one of the main components of reserves.

As a consequence, we are now in some kind of a no-man's land. This is so true that, at the present session, our American friends have been saying that

[1] The conclusions and evaluations of the Session on National and Supranational Monetary Authorities' Position of the World Conference on Gold were drawn, at the end of the Conference, by René Larre, Chairman of the Session, while a final comment was made by Peter Oppenheimer, opening speaker of the Session.

they have no gold policy and the European Commission representative went so far as to express doubts as to the future gold policy of the EEC.

2. The Present Situation: High Freedom of Monetary Authorities.

Meanwhile, in the absence of political discussion on the role of gold, monetary authorities have enjoyed a high degree of freedom, a degree they had not known for several years.

For the time being, for instance, they may buy and sell gold on the market, which they were forbidden to do for two years after 1976; they can value their stocks at any price they choose, between $35 — some of them do — and $800 (some of them may have done so). They may borrow against gold, using the metal as a collateral. They can change the composition of the reserves; they can do whatever they like provided they respect two limitations. One is that there is no fixed price for gold; as a consequence, it is very difficult to make out right sales or purchases, because, the day after, the wisdom of your decision may be denied by the market. This explains the preference for swaps and other reversible operations.

There is another proviso: that monetary authorities have to be aware that public opinion is very sensitive to the national gold stock. In every country gold is now considered as part of the national heritage and it has become dangerous for governments and governors alike to tamper with the gold reserve of the nation.

But except for those two limitations, monetary authorities have a field day.

3. The European Monetary System

One of the consequences of their freedom has been the institution of the European Monetary System (EMS). Even if there are some doubts as to its merits, in the mind of the Commission, the member governments seem to be very keen on this scheme as they find it a very convenient way to mobilize their gold and, at the same time, to upgrade its book-value. We heard at the dinner last night some confirmation of this situation, from the Italian authorities. So the EMS seems to be in business for a long time and it may be considered as the response of the Europeans to the OPEC policies.

4. A Glance to the Future: No Remonetization of Gold and Central Banks' Intervention

While we have been talking mainly about the past and the present, we have been touching a little bit on the future.

From what we have heard during the session, there is no strong feeling in favour of remonetization of gold. Even the word has not been used. I confess that it came to me as a surprise that nobody took advantage of this meeting to advocate a return to the gold standard. Whatever may be the other reasons, one seems to be that the level of inflation discourages even the boldest supporters of gold.

There is not much scope, either, for central bank interventions. Even if political restrictions have been removed, there is no real expectation central banks would step back into the market. After all, Governors do not like to act alone and, at the present time, there is no agreement among them to embark on a joint action. In this connection, we have heard the statement of the members of the Congressional Gold Commission. We were told that the report is expected to be released in the coming weeks, but that it will not be a milestone in the field of the American gold policy.

Reviewing this situation, one is led to the conclusion that the market will be left to its own forces and devices during the foreseeable future. Some central banks will sell gold, but they will be the usual sellers: South Africa, Canada, the Soviet Union. For them it will be business as usual, as they are the normal suppliers of the market.

On the other hand, some central banks will buy, like the surplus OPEC countries and maybe Japan, but it will be a matter of individual purchases.

Should a larger demand from monetary authorities develop, it would be a windfall for the market as — as it has been said by various speakers — the demand from industry, jewellery and private investment may be rather weak. This is a somewhat disappointing conclusion, but it is the conclusion that emerges from our discussions in session.

However, the situation would be different in two cases: one would be a threat of war. In that event interest for gold, as in the past, would be revived, but this is not a very pleasant prospect. The other case would be runaway inflation, but we have been told by our economic forecasters that we have not to be afraid of a resurgence of inflation.

So we cannot escape our undramatic conclusion.

PETER M. OPPENHEIMER

5. "It's an Ill Wind that Blows Nobody Any Good"

We have had a long morning and I think I ought not to prolong it still further; I shall keep my comments as short as I can, consistent with fulfilling my responsibility to contribute a fair share of the concluding remarks.

We have a saying in English: "It's an ill wind that blows nobody any good." In other words, it is unusual for any unfortunate circumstance or disaster to harm absolutely everybody and benefit nobody. Accordingly, I have found it amusing to meet at this Conference all the excellent traders, manufacturers, bankers and others in the gold market with a vested interest in instability, volatility, uncertainty, anxiety and inflation.

I appreciate, moreover, the point made by Miss Du Boulay and others that gold investors in South Africa and elsewhere are fundamentally conservative people; they have to take decisions which commit very large amounts of capital and they have to proceed cautiously. So their willingness to commit capital in recent years to new gold production and to commercial activity in gold indicates their confidence that instability, volatility, uncertainty, anxiety and inflation are here to stay.

This might appear to create an ineluctable conflict of the sort propounded by Professor Aliber between commodity interests in gold and monetary interests. I don't myself accept that any such fundamental conflict is inevitable. I therefore answered both Aliber's questions in the affirmative. Of course, there may be differences of interest at a superficial level in the face of particular events; but the whole point about the monetary role which gold has played in the past, including the Bretton Woods system before 1971, is that it could not have played that role unless it (gold) was a commodity. The breakdown of Bretton Woods marks the watershed between a managed but still gold-based system and the pure-credit-money system which followed and which makes floating exchange rates unavoidable.

Previous speakers this morning have reflected the widespread consensus that there is not going to be an early return to a gold-based system. And by this term I don't mean a 19th-century-style gold standard (about which I shall say a word before I finish), but simply something like the Bretton Woods system, with a gold basis to international currency arrangements. On the other hand, I am not sure how much consensus there is as to whether a return to gold, if feasible, would be desirable. Even among monetary authorities, governments, politicians and professors of economics, I don't know whether there is adequate appreciation of how much the departure from gold, as a result of the refusal of the US authorities to raise the price in a controlled and deliberate way in the 1960's, has actually cost us in terms of lost stability.

Perhaps it is too much to ask that there should be a close consensus on this, because the question is actually a very tricky one. There is scope for not one but several research programs to investigate the question of how much of the inflationary outburst of the early 1970's, and the shift in the terms of trade and income distribution at that time, were in some sense a consequence of going off gold. Clearly, it was not all due to that: but, equally clearly, some of it was. The mushrooming of dollar reserves in those years created a permanent acceleration of inflation, with whose consequences we are still living.

6. Inconclusive Attitudes of Official Bodies on the Use of Gold

If I may now briefly supplement Mr. Larre's remarks of a few moments ago, what emerged very plainly from our part of the Conference was the piecemeal and inconclusive attitudes of official bodies and governments on the steps that might be taken to try to increase somewhat the use made of gold in the present monetary system. This is, of course, against the general background that gold is a relatively immobile component of reserves, sitting for the time being at the bottom of the pile. Getting it mobilized is a great problem, when there is no price certainty underlying the market.

On the other hand, the steam has gone out of efforts to remove gold totally from the monetary system. I should be surprised if large-scale sales of gold by the US authorities were to be resumed. I agree with Miss du Boulay on this. Similarly, the International Monetary Fund, although perhaps a little puzzled as to what it is supposed to do with its gold and whether it is now supposed actually to love the metal, has at least quieted down after the hate campaign against gold which it orchestrated for much of the 1970's. There is unlikely to be a resumption of the Fund's drive to sell off the officially held gold to the private market in the years ahead.

But, apart from these reflections, there is little to be said. The European Monetary System has run into an impasse. It has taken a first step towards mobilizing gold indirectly through the ECU mechanism, but no second or third steps are in sight. The IMF, as Dr. Wittich explained to us, now professes to see the value of its gold holdings, but isn't actually going to do much with them. Henry Wallich emphasized that current US official opinion favours keeping gold at the bottom of the pile. The same sort of conclusion came out of Rayner Masera's paper about the returns attained in a commercial sense by central banks on their gold holdings; and I think also out of the interesting technical proposal put forward by Jürg Niehans for a short term mobilization of gold as an instrument of monetary management. Niehans himself stated that he didn't expect early acceptance of this proposal; and even if it were accepted, it would imply only rather tentative excursions into the gold market by individual central banks on an *ad hoc* basis. It would not imply a system-wide rehabilitation of the metal as a basis for international monetary arrangements.

7. Two Key Factors: Politics and Inflation

Underlying these inconclusive attitudes are two main factors. The first is political preference. Even though the hate-gold syndrome plays a much smaller part in political attitudes now than it did a few years ago, there is still the fact of South Africa and the Soviet Union being the major suppliers of gold, and the demagogic question-marks over their respectability in the family of nations. There is also still an undercurrent in some circles, not least the academic, which feels it intellectually antiquated and a confession of failure to admit that gold could still have a positive monetary role.

The second underlying factor, and probably the more important one, is the problem of inflation. In his delightfully drafted paper, Niehans described the current revival of sentimental support for the pure 19th-century gold standard as the financial equivalent of creationism in biology. I think that this absolutely hits the nail on the head. It would be so nice if we could revert to some kind of primitive medicine-man magic — economists being the modern witch-doctors — which would enable governments to take a convenient academic pill and thereby squeeze inflation painlessly out of the system. (Please excuse the mixed metaphors.) Such magic simply isn't available. Monetary authorities, having learned the power of controlling money supplies independently of any outside constraint and relying only on their judgement, their technical competence and the democratic political pressures to which they are subject, cannot then go back and say, "Alas, this was all a mistake, and what we really should do is tie one hand behind our back and then proclaim that we are unable to perform handstands because we have only one hand available."

What is necessary is to come to grips with the problem of inflation in today's context with today's weapons, accepting that it is partly a political problem and that, to the extent that it remains a technical-monetary problem, it has to be handled by the modern sophisticated methods of central bank and Treasury management and not by trying to re-erect the golden totem.

Should it happen that, recognizing this, monetary authorities do indeed succeed over a run of years in bringing inflation down again to low levels, without major long term loss of employment and economic activity, it would then be possible, should governments be so minded, and notwithstanding anything that the commercial sector in gold might be doing, to re-establish a Bretton Woods-type gold-based system. This in turn would provide a valuable set of guidelines and benchmarks for policy, helping governments to maintain national stability thereafter. But the initial breakthrough against inflation is a prerequisite for that. It cannot itself be achieved by "going back to gold".

Yet, once the breakthrough had been achieved, would governments and central banks still think it worthwhile to re-adopt gold? That is a further and crucial question; because, after all, having proved that financial stability could be regained without gold, why should they choose to bring gold back into the picture thereafter? The pre-eminence of national sovereignty and the limits on international cooperation provide good reasons for doing so; but one may doubt when the governments will have the insight to acknowledge this in practice.

So, at the end of the day, if I am obliged to make a forecast, I too think that the commercial sector in gold will continue to rule the roost and will not find officialdom trying to resume control of the market in the foreseeable future.

INDEX OF NAMES

Aliber, R.Z., 9, 49, 151-161, 171, 173, 179, 181, 284

Baffi, P., xx
Balestra, G., 138
Bentsik, E., vii
Bernstein, E.M., 223
Bignardi, F., vii, xix
Boni, P., vii
Bordo, M.D., 49
Boyer De La Giroday, F., 14, 205-226
Braudel, F., 205
Branson, W., 254
Brunelleschi, F., 140

Cacace, N., vii
Cairnes, J.E., 2
Calabi, N., 1
Cantillon, R., 2
Cassel, G., 3
Cellini, B., 140
Cesarini, F., vii
Cirri, G., vii
Coyne, H.J., 79, 80, 111-124, 178, 179, 180, 263, 268

Dante, 74
Davies, J.L., 50
Dealtry, M., 197, 206
De Blasio, P., vii
De Gaulle, C., 200
Del Castillo, B., vii
De Vincolis, A., vii
Dini, L., 15, 75-78, 79
Dornbusch, R., 254
Du Boulay, L., 78, 85-100, 173-181, 284, 285

Ehrlicher, W., 273

Falchi, G., 10, 15, 245-262
Fells, P.D., 78, 85-100
Fisher, I., 3
Friedman, M., 195, 207

Galiani, F., 2, 13
Gallo, P., vii
Gazmararian, D., 63-68, 75, 79
Ghergo, A., vii
Ghiberti, L., 140
Gianani, F., vii
Giangrossi, L., vii
Gilbert, M., 4, 193, 197, 198, 199, 200, 201, 202, 206
Gilbert, P.A., 130, 135
Gori, V., 138
Grassini, F.A., vii
Gray, R.W., 50
Green, T., 59-62, 75, 80, 178
Guy, R., 69-74, 77

Haarman, L., 1
Hanselmann, G.R., 21-36, 75, 76, 78-81, 174, 179
Harrod, R.F., 3, 4
Hawtrey, R., 2
Hicks, J., 163
Higonnet, R., 168
Hinshaw, R., 4, 5, 13
Hirsch, F., 206
Hume, D., 2, 4

Jastram, R., 5, 6, 14, 99, 143-150, 179, 181
Jeanty, P., 37-42, 79
Jevons, W.S., 2
Johnson, L.L., 50

Kaldor, N., 163
Kalecki, M., 163
Kantarelis, D., 125
Kendall, C., 151
Keynes, J.M., 2, 3, 4, 168, 205, 216, 245
Kouri, P., 254

Laffer, A., 104
Lamborghini, B., vii
Languetin, P., 171-173
Larre, R., 281-283, 285
Leutwiler, F., 73
Locke, J., 2
Lombardini, S., vii
Longo, A., vii
Loraschi, G.C., vii

Main, T., 85, 101-109, 174
Malthus, T.R., 2
Marsh, D., 223
Marshall, A., 3
Marx, K., 2
Masera, F., vii, xx
Masera, R.S., 10, 15, 245-262, 265, 285
McDonald, G., 128, 135
Meade, J.E., 272
Medici, G., vii
Merloni, V., vii
Molière, 81
Montesquieu, 81
Morgenthau, H. Jr., 272
Mucci, A., vii
Mun, T., 2
Mundell, R.A., 272

Nesi, N., vii, xix
Nessim, R., 43-50, 76, 78
Niehans, J., 14, 15, 80, 271-279, 285, 286
Nurkse, R., 195

Oppenheimer, P.M., 14, 78, 79, 187-203, 206, 281, 283-287
Orlandi, F., vii

Pantaleoni, M., 3
Paoluci, S., vii
Pareto, V.F.D., 3
Parrillo, F., vii
Pasqua, G., vii
Petty, W., 2
Plass, F., 51-57, 77, 79
Polak, J.J., 205, 206, 216
Pollaiolo, A. del, 140
Ponzellini, M., vii
Potts, D., 127, 135
Prodi, R., vii, xix-xx

Quadrio-Curzio, A., vii, xx, 1-17, 77

Ravenna, R., vii
Reagan, R., 30, 44, 130, 133, 134, 183, 242
Renard, J.P., 81
Ricardo, D., 2, 4, 8, 164
Ricci, G., vii
Richardson, J.D., 50
Richter, R., 273
Robbins, L.C., 4, 5, 13
Robinson, J., 163
Roosevelt, T., 272
Rueff, J.L., 200, 206
Ruffolo, G., vii

Sacchi, C., vii
Schlesinger, H., 223
Schwartz, A., 5, 6, 7, 14, 182, 237-243, 263, 264, 265, 268
Smith, A., 2
Solnik, B.H., 128, 135, 254
Solomon, R., 223
Spolverini, A., vii
Stella, L., 80, 137-142
Sylos-Labini, P., 15, 163-170, 181
Szego, G., 254

Tantazzi, A., vii
Tinbergen, J., 272
Torboli, F., 80, 137-142, 176, 177
Triffin, R., 4, 50, 192, 198, 199, 205, 208, 216, 228, 267

Van Eck, J., 127, 135
Van Tassel, R.C., 9, 125-135, 179, 180
Verrocchio, A. del, 140

Wallich, H., 14, 223, 263-269, 271, 285
Walras, L., 3
Webb, S., 223, 224
Webb, T., 205
Wicksell, K.J.G., 3
Wieser, F. von, 3
Wittich, G., 176, 227-236, 285

Zijlstra, J., 73, 206, 223
Zucchi, P., 138